The UFO Technology Handbook

for Saving the Environment and Exploring the Universe

by **Paul Hynek**

and Lamont Wood

Printed in the United States of America

ISBN 978-1-953321-07-7

UFO Books Press
A division of Micro Publishing Media, Inc
PO Box 1522
Stockbridge, MA 01262

www.ufobookspress.com

Table of Contents

Introduction:
If They Can Do It, We Can

About the time this was written (in the summer of 2020) a member of the US Senate was demanding assurances from the Pentagon that the unidentified flying objects that had been filmed over certain American military bases were not of this Earth. Considering the technological leap over American hardware that they appeared to embody, he indicated a preference that they were operated by extraterrestrials rather than by national adversaries (i.e., fellow humans.)

And indeed, our visitors (as we'll call them) have long exhibited feats of technology that we can't touch, conducting both sustained hovering and high-speed aerial movement while exhibiting neither fuel tanks nor heat trails. And the fact that they are clearly not of this Earth indicates they possess some form of interstellar travel, which we also cannot now match. (Alternately, as will be discussed, they may also possess some form of time travel, which we likewise cannot now match.)

But if we cannot now match the technology, the premise of this book is that our inferiority is a temporary phenomenon. We have made such enormous strides since the start of the industrial revolution (roughly 1750) that we are now at the point where we can offer some level of informed speculation about how we might someday match what we see.

And it is not faith or wishful thinking that leads us to make such predictions, but the use of the only rational method we have for predicting the future: extrapolating from the past. And our recent past has been marked by the discovery that scientific and technological innovation are key national resources. These resources are unlikely to be abandoned.

But even simple faith would suffice for these predictions. It may be one of those stories that are too good to be true, but early in the industrial revolution a politician supposedly once asked British scientist Michael Faraday (1791-1867) if this new thing he kept talking about, called "electricity," honestly had any practical use. (Presumably this was in the late 1820s, when electricity was still the province of desktop demos and parlor tricks, and before the telegraph was invented, establishing electricity's usefulness.)

Faraday's reply: "One day milord will tax it."

And now, two and a half centuries later, we have a huge electrical power industry, and it generates tax revenue. Electrification, meanwhile, allowed a second industrial revolution, and then the subsequent computer revolution. Technologies, meanwhile, built on each other in unexpected ways: industrial scale extraction of aluminum would not be possible, for instance, without industrial-scale electrical generation, since aluminum is extracted from its ore via electrolysis rather than smelting. Meanwhile, without the large-scale availability of aluminum today's aeronautics industry would not be possible. Electrical appliances and lighting have meanwhile allowed the humblest citizens to enjoy luxuries and conveniences previously possible only for those with a large staff of servants— what do you think it took to enjoy a hot shower in Faraday's time?

Faraday, you'll note, actually made his prediction on faith, as he had no extensive history of the industrial revolution to fall back on. But he was familiar with human greed, competitiveness, and ambition, and had seen how those traits interact with innovation. But he also expressed faith in electricity without ever having seen a hydroelectric dam, or a city brilliantly lit at night, or an e-commerce screen. He didn't need such details.

We, on the other hand, can make predictions based on leading edge science and speculative theories whose supporting math was rendered with cybernetic precision. And we can be confident that humanity will continue to strive to grasp what it has glimpsed, and turn theory into practice, ultimately bettering the condition of the Earth's inhabitants. (Admittedly, new technology does not automatically lead to better conditions—consider the huge sums spent on weapons' development. But new technology invariably gives us more options.)

Beside glimpsing future possibilities through the non-physical beauty of mathematical symmetries, there are also the beauty of physical things glimpsed in the skies by countless witnesses. Whoever operates them lives in the same universe we do, and therefore are ruled by the same laws of physics. Once we master those laws, we can expect to match the visitors' feats.

As a side benefit to our striving, we could expect to master things like immaculate energy, which should allow us to enjoy electrical power without strangling ourselves with pollution; and cheap access to space, letting us not only become a space-faring civilization but develop technologies (if not whole industries) that are not practical on the Earth's surface, beset as it is by gravity and air pressure.

While it's often said that sufficiently advanced technology is indistinguishable from magic, this book is not about fantasies— everything described could, in theory, be achieved, if certain theories can be scientifically confirmed and then reduced to practical technology. We don't pretend that every suggestion in the book can or will be achieved. But, as in tennis, you miss all the shots you don't take. And successfully pursuing even one of our suggestions would, we believe, be of enormous benefit.

Nor do we pretend that any of these things can be achieved over-night. But remember Faraday, who made his successful prediction 200 years ago knowing far less about science and technology than we do. Considering what could be done in another 200 years, the mind boggles. Nor is there any deadline—in 500 years, 1,000 years, or 10,000 years, there's every reason to assume that we can match any technology we see on display.

And we're also pretty sure that someone will figure out how to tax it.

Meanwhile, throughout the book, when discussing energy, we will typically use the joule from the International System of Units as a basis for comparison. A joule amounts to one watt per second, or, mechanically, the force of one newton expressed for one meter. (A newton is the acceleration of one kilogram at one meter per second squared. Ironically, considering the association between Isaac Newton and falling apples, a newton is also about the force your hand exerts against gravity while holding the average apple.) We will also use the AA and AAA battery (often used in electronic devices) as a basis for comparison. Off the shelf, the long-life alkaline version of the AA battery should store 9,360 joules, and the smaller AAA version should store 5,071 joules. (Other chemistries may store dramatically less energy, and in all cases their shelf lives are limited.)

We will use simplified exponents to render large numbers. For instance, 9,720,000,000 (nine billion seven hundred twenty million) would be rendered as 9.72E9.

❖   ❖   ❖

1

Advanced Energy: Overview

Becoming a spacefaring civilization and matching what we see in the sky will require energy—lots of it. That may not sound like a positive statement, since (in global terms) supply can hardly keep up with demand now, but there are other sources of energy we could tap in the foreseeable future that could dramatically expand humanity's energy supply. Oil wells may someday go dry, but there's not going to be any shortage of solar energy—in fact, the available amounts could be mind-boggling. Meanwhile, matter/antimatter fuel puts the energy density of a hydrogen bomb in the shade, and why even bother with fuel when you can just drain the heat out of any liquid you have handy?

New technology and physical plant may be required—huge amounts of it in some cases, dwarfing hydroelectric projects. The nature of some of the technology cannot be currently envisioned. And there will be physical dangers.

But that's always been the case, and it has not stopped us yet. As noted in the Introduction, British scientist Michael Faraday in the 1820s could not have envisioned cities ablaze at night with electric lights—in fact he'd never seen a lightbulb—but he was still confident of one fundamental prediction: electricity would be a valuable resource in the future.

The following predictions are based on more evidence than Faraday had. We'll see how they turn out.

❖   ❖   ❖

1.1 – Advanced Energy:
Matter/Antimatter Annihilation

Matter is composed of atoms, and those atoms are composed of subatomic particles such as protons and electrons. Each of those subatomic particles has an electric charge, and that charge is standard in all matter we've encountered: protons have a positive charge and electrons have a negative charge, etc. There appears to be no reason for the polarity of these charges, and atoms that have suffered powerful collisions, such as in nuclear accelerators, or in the atmosphere by cosmic rays, will sometimes end up with opposite charges on their particles—i.e., they'll become antimatter.

Beyond the electrical polarity of the charges, there is no other difference between matter and antimatter. Compared with matter, the mass and other physical properties of antimatter is the same. Chemical and nuclear reactions that antimatter atoms have with each other will have the same outcomes as ordinary atoms have with each other. (Despite the name, however, antimatter does not generate antigravity.)

But there is one major issue with antimatter that makes it compelling in terms of advanced energy and space travel: when matter and antimatter are brought into contact, they will annihilate each other and their mutual masses will be entirely converted into some form of energy, in the amount prescribed by Einstein's famous equation $E=mc^2$. No intermediate processing or nuclear reactor will be necessary. The process will be 100 percent efficient in terms of converting mass to energy, while even an H-bomb is no more than one percent efficient in converting mass to energy.

The result is a startling amount of energy. A half-gram of matter (the size of a raindrop) added to a half gram of antimatter will produce a gram that will be totally and instantly converted to energy (mostly heat.) The resulting blast should equal a 21.5 kiloton bomb, or one with about 50 percent more power than the one used on Hiroshima. If you combined the two droplets in your kitchen using matter and antimatter eyedroppers, your neighborhood would be replaced by a smoking crater at least a hundred yards in diameter and every building would be flattened for at least a mile in all directions. But there would be no radioactive fallout, as with a nuclear explosion, or even toxic fumes, as with a chemical explosion, since all the matter would be converted to energy. (There'd still be dust from the blast and smoke from the flaming debris, of course.)

Obviously, with a source of power like that, our energy problems would be gone. As a portable energy source, in terms of energy density antimatter has no equal. The main challenge would be, as usual, to not kill ourselves with it. But there may not be much danger of that anytime soon, as there are two huge drawbacks when considering any employment of antimatter.

The first is that we don't have a significant source of antimatter, and we can barely say we have an insignificant source. While the difference between matter and antimatter seems so arbitrary that you'd expect that half the universe would be antimatter, in fact we see no antimatter anywhere we look. If an antimatter celestial body was invading space dominated by ordinary matter (or the reverse) we'd see a halo of annihilating particles around it. Astronomers have detected no such event. Perhaps matter got the upper hand over antimatter in the first moments after the Big Bang, making it a really big bang. Or perhaps there are distant neighborhoods of the universe, not visible to us, that are reserved for antimatter. We just don't know.

But antimatter does exist on Earth and we do have access to it— in quantities so small it makes dust motes look like aircraft carriers. The only regular source of antimatter is nuclear accelerators, which create small amounts as a byproduct. And by small amounts, they mean it would take centuries to make a nanogram (equal to the size of the average human body cell.)

Of course, there was a time when we had no sources of refined gasoline, and now we have an economy founded on the personal, gasoline-powered automobile. Presumably, specialized nuclear accelerators could be developed that could economically generate antimatter in perceptible quantities, assuming the developers had sufficient financial motivation.

But that motivation is not likely to be forthcoming until antimatter's second big drawback is overcome: storage. Any container we construct for it would be made of ordinary matter, so putting antimatter in it would trigger an explosion. The only apparent solution would be to trap it in some kind of magnetic suspension. You'd want the suspension to be reliable for obvious reasons, but you'd also want the suspension to last long enough yet be agile enough to transport the antimatter to whatever matter-annihilation chamber is in use, and feed it into that chamber under total control.

Very likely antimatter production and antimatter handling technologies will advance in parallel, much as (about 250 years ago) the development of steam engines and of machine tools depended on each other. Currently, it would cost the equivalent of humanity's entire economic product for several decades to produce one gram of antimatter, but, presumably the price will start coming down by orders of magnitude while handling technology likewise improves by orders of magnitude.

Presumably, we'll eventually start industrial-scale electrical plants that would use heat generated by matter-antimatter annihilation to make steam to run turbines. Tiny amounts of antimatter with limited storage would suffice.

But matter-antimatter power stations would have to be able to generate more electricity than it takes to produce the antimatter they consume. That's probably possible considering that antimatter is only half of the mass of the matter that has to be annihilated to make energy, the other half being regular matter, which we don't have to produce.

It would probably be enormously helpful to make antimatter in orbital factories powered by solar energy. Cooling and isolation of the antimatter ought to be easier there, and the electricity to power the process will be free, once enough solar power collectors are built. Also, accidental explosions won't transform inhabited landscapes into cratered wastelands. It is also possible we will be able to mine antimatter in outer space, scooping it up in places where gas is exposed to collisions with subatomic particles, such as in Earth's two Van Allen Belts, or in the upper atmospheres of the gas planets.

But again, we face chicken-and-egg problems: there are all sorts of things we could do easier in space, if we could get there readily.

But regardless of how we do it, once we start making antimatter in quantities that make it worth transporting, then we can talk about it for space travel, as laid in in Section 4.4.

If visitors were using antimatter as a form of fuel, our main clue would be the absence of any other obvious fuel source for their spacecraft. The fuel-to-payload ratios allowed by antimatter rocketry more closely resemble over-the-road trucking than conventional rocketry using chemical fuels. Interestingly, with many visitors, there in fact is no obvious fuel source.

But keep in mind that antimatter rocketry would still be rocketry, confining its users with the same straitjacket of Newtonian laws that's laid out in Section 4.0. Antimatter technology would be compellingly useful for local excursions, but of no particular advantage for interstellar travel. For that, you'd need something other than rocketry. We'll be exploring possible alternatives in Sections 6.x.

❖   ❖   ❖

1.2 – Advanced Energy: Advanced Solar

The sun amounts to a nuclear fusion reactor that we did not have to pay to build and do not have to pay to keep fueled and maintained, which bathes us with a generous amount of energy that we can use for free. But once you get past the preceding sentence solar power sounds less and less idyllic, due to practical problems introduced by issues that include atmospherics, geography, the laws of thermodynamics, and a periodic cutoff of the source called "night."

When used in space, many of these disadvantages go away, leaving us with the big one: you have to get into space in order to use the advantages offered by space. When used for propulsion, solar power offers you a form of rocketry that uses minimal fuel—or none at all, if you unfurl your solar sails. But ultimately solar propulsion involves the same physics as rocketry (albeit with very different numbers) which means that using it for convenient interstellar travel is not likely.

As for that free bath of energy, a surface of one square meter perpendicular to the rays of the sun, at Earth's distance from the sun, receives 1,362 watts of energy from the sun. So there should be no problem powering a house in the suburbs (which draws an average of 2,000 watts) with a solar collector of only two square meters (about half the area of US-standard king-sized bed)—if your collector is 100 percent efficient, and is positioned outside Earth's atmosphere.

That's never the case, of course. About 30 percent of the radiation is lost in the atmosphere before it reaches the Earth's surface, and there are further losses according to latitude, weather, and nightfall. As for efficiency, solid-state photovoltaic (PV) power converters, preferred for their lack of moving parts, have efficiencies that appear to average around 20 percent. (They also have a working life of about 30 years, similar to the roofs they are often installed atop.) The trend in PV technology is toward greater efficiencies, and average efficiency may reach 40 percent in the foreseeable future. (The technology started out at two percent in the 1950s.) Large-scale solar installations may involve sunlight concentrated by mirrors and lenses to supply heat for generator turbines, replacing fossil fuel as a source of heat. Such turbines have efficiencies in the range of 35 percent.

As an immaculate energy source, solar would appear unbeatable—any shade-giving structure's roof could be used to generate power until its PV panels age-out. But solar-rich areas (i.e., deserts) rarely coincide with heavy usage areas (i.e., cities), and we can't forget those factors of weather, efficiency, and especially night. In the end, unless the power grid can be made transoceanic (so that the kilowatt-giving sun never sets on it) solar power cannot be our sole reliance.

We might get to a night-free grid faster by using solar collectors in geostationary orbit, which pass only briefly through Earth's shadow. They would convert unfiltered sunlight into microwaves that would be beamed down to ground collectors, and there converted into electricity for the grid. This would require in-space construction of an edifice with hundreds of acres of PV collector arrays plus microwave antennas, whose total mass would approach that of an aircraft carrier. Hundreds of space launches would be required. Pundits assure us that such a project would be economically competitive with conventional terrestrial power sources if launch costs were about one percent what they are today (i.e., $100 per pound, as opposed to $10,000 per pound.)

But if you are going to be in space anyway—hang the cost—then solar power could serve the needs of electric rockets. As will be laid out in detail in Section 4.0, rockets move by expelling "reaction mass," and the higher the "exhaust velocity" of the reaction mass, the better. With conventional chemical rockets, the exhaust velocity typically peaks out at about 2.7 miles per second. With electric units you can get exhaust velocities closer to 60 miles per second. But what you won't get is thrust—electric units cannot lift cargo into orbit. But once there they can continue accelerating for long periods, since they're using small amounts of propellant—if they have sufficient electricity. For supplying that electricity, solar power collectors are an obvious answer.

It's also possible to skip power conversion completely and use the raw force of sunlight like a sailboat uses the wind (sort-of, as you won't be able to tack into the sunlight like you can tack into the wind.) At Earth's distance from the sun, sunlight exerts a pressure of 9.08 one-millionths of a newton per square meter on a perfectly reflective surface pointed directly at the sun. A newton is the force needed to accelerate one kilogram at the rate of one meter per second per second, so a sail of one million square meters (one square kilometer, or about half the size of the average American farm) would feel a push of 9.08 kilograms per second per second. If the sail was made out of aluminum foil from your kitchen it would weigh 432 metric tons. Let's assume dramatically thinner foil, giving a mass of 100 metric tons. It would experience a push of 0.09 mm per second per second. The push would add up second by second, so that in a little more than an hour and a half you'd have accrued enough velocity to leave Earth orbit, all without using any fuel (overlooking the nontrivial task of launching and erecting the sail, etc.)

Except if you just start sailing away with the sunbeams, you'll probably not get anywhere. Space travel is rarely about blasting from points A to B—typically, you add nudges to your existing orbit to alter it so that you intercept your target. With a solar sail, you would want to use it when the pressure of sunlight would alter your orbit in the desired direction, and that would only be a fraction of the time. The rest of the time you'd want to turn you sail sideways to the sun, effectively furling it, so that it does not pull you off course or perhaps even decelerate you out of orbit. It might take multiple orbits to acquire the desired acceleration in the desired direction, amidst complex maneuvers and calculations.

For interplanetary travel the situation only gets worse, as the pressure of sunlight falls as you get farther from the sun, by the square of the distance. For instance, the asteroid belt is about three times farther from the sun than the Earth, so by the time you got there the pressure of sunlight would be one-ninth what you experienced in Earth orbit.

So solar sails, like sailboats, would seem more suitable for those who are in no hurry. In any event, solar sailing involves Newtonian physics (even if no fuel is expended) meaning that exceeding 10 percent the speed of light is unlikely. If you wanted to spread you sail and head away from the sun in some random direction, you could leave the solar system and in the course of many decades or centuries reach another one. But without ongoing maintenance your gossamer sail would probably be so damaged by debris encountered in the interstellar medium that being able to decelerate in order to stay at your destination would be unlikely.

So if a sail-driven craft did arrive in our solar system from the outside, we can be sure it has a crew, live or robotic, who performed maintenance during the trip. We'll need to prepare a welcome, peaceful or otherwise, based on that assumption. There'll be no hurry as the craft will have to perform convoluted maneuvers to decelerate into a useful orbit around our sun, potentially taking years. (An uncrewed sail-craft from any but the nearest star would probably arrive unnoticed, since it would just be scattered debris.)

If we then managed to talk to them, and bring up terrestrial uses of solar power, even including satellite collectors, they might think we're making baby-talk, as we'll discuss next.

❖ ❖ ❖

1.3 - Advanced Energy: Dyson Engineering

As laid out in the previous section, with cheap access to orbit it would be possible to build solar collectors in geostationary orbit that beam power down to the Earth. Such energy would be free, but you'd still need to launch hundreds of cargos into orbit to set up the power stations, and you'd still need a vast terrestrial power grid to distribute the newly acquired energy, etc.

But why stop there? If cheap access to orbit (i.e., about one percent the current cost of roughly $10,000 per pound) is the norm, why not fill the space around the Earth with more and more power stations, presumably beaming power to each other and then to the original geostationary satellites, and then down to Earth?

But why stop even there? You could fill Earth's orbit around the sun with power stations that pass energy to each other around the orbital path back to Earth. After that you could fill the space around the sun with orbiting power stations in every available orbital plane, carefully positioned and scheduled so as to avoid collisions. For that matter, you could enclose the sun in a thin shell that the Earth's orbit could fit into, capturing the entire output of the sun, and increasing the level of solar energy available to humanity by factors of trillions.

Such schemes are referred to as Dyson engineering, after Fellow of the Royal Society theoretical physicist Freeman Dyson (1923-2020) who first popularized the idea of such constructions in a 1960 paper. He was undertaking a thought experiment to explore the idea that humanity's increasing thirst for energy has no bounds, and would eventually require the total output of the sun. That output could be captured for humanity's benefit by enclosing the sun in the previously described shell, sometimes called a Dyson Sphere.

Other writers have taken the Dyson Sphere idea and run with it, but Dyson never said it was practical—and it clearly isn't. Such a shell would require most of the solid matter in the solar system, and the components could not be positioned without them falling into the sun. Even if somehow constructed the sphere would surely shatter under the enormous stress it would experience.

But the less fervid offshoots of Dyson's idea are not automatically impractical, such as lining Earth's orbit with large solar power stations, called a Dyson Swarm. But practical or not, such ideas remain utterly uneconomical, as the era of cheap access to orbit has still not dawned.

But such may not be the case everywhere, which brings us to Tabby's Star, otherwise known as KIC 8462852, or that 11th magnitude star (i.e., far too faint to be seen with the unaided eye) in the constellation Cygnus, at right ascension 20 hours 6 minutes 15.4527 seconds, declination 44 degrees 27 arcminutes 24.791 arcseconds north. From the ground, it's just a couple of finger-widths separation from the star Deneb at the top of the Northern Cross.

In 2015, a team of citizen astronomers under the name Planet Hunters, led by astronomer Tabetha S. Boyajian, were examining data from the Kepler space telescope and found that Tabby's Star had odd periodic swings of brightness of up to 22 percent.

That's pretty extreme variability, especially for a star of mature age. Plenty of theories have been developed to explain the swings, usually involving debris or objects orbiting the star. These theories also include large artificial constructions such as the sun-orbiting power stations suggested by Dyson, presumably built for the reasons he suggested. So, yes, under one theory the variability of Tabby's Star could be evidence of an extraterrestrial technical civilization.

The star is about 1,470 light years away, so conclusive surveillance is not possible with current technology. Meanwhile, some researchers say that the nature of the dimming of Tabby's Star is not consistent with the blocking of light by solid objects. (Of course, if they were Dyson's power stations, they'd be predominately latticework rather than solid objects.)

Actually, we might be better off hoping for a natural explanation. If we are seeing monumental orbital constructions by the Tabbyites, it would imply that a technical civilization that is clearly centuries ahead of us has given up on interstellar expansion and has committed to looking inward, thoroughly exploiting their own star while turning their backs on the countess stars around them. This could only mean that they reject interstellar travel. Perhaps they did it for cultural reasons—they're xenophobic. Or, perhaps they found interstellar travel to be too difficult. Perhaps they've even decided it's impossible. That's not an example that we need to see.

Alternately, we may be seeing an example of what's called the transcension hypothesis, which states that an advanced civilization is more likely to exploit inner space than outer space, as the process of scaling down is more efficient and controllable than space exploration. If they pursue it to the black hole level they will disappear from our universe (i.e., into one they created, presumably to their precise specifications, inside a black hole.)

(Incidentally, man-made black holes are entirely possible, as laid out in Section 5.6. They do require a lot of energy, so a Dyson Swarm would be handy.)

But whether the Tabbyites are outbound or inbound, if those genuinely are monumental artificial constructions orbiting Tabby's Star the Tabbyites have clearly achieved cheap access to orbit—a heartening example.

Meanwhile, if some other extraterrestrial civilization had somehow managed to build a true Dyson Sphere, we might not ever know about it, since there'd be no escaping light for us to detect. We might be able to spot one that's under construction, although there's some question as to what that would look like—Tabby's Star?

There are, incidentally, other stars with odd variability that are under investigation. Whether specifically the Tabbyites are real or not, Dyson may yet get his validation.

❖   ❖   ❖

1.4 – Advanced Energy: Kardashev Engineering

If Dyson engineering (outlined in the previous section) makes our current level of technology look like baby stuff, then Kardashev engineering makes Dyson engineering look like baby stuff. In 1963 Soviet (later Russian) astrophysicist Nikolai S. Kardashev (1932-2019) came out with a scale for rating the technological standing of a civilization based on the amount of energy its denizens are able to use. (What they use it for was not addressed, but history indicates that when energy is available, some use will be found for it.) His original scale allowed for three levels, or types of advanced civilizations.

Type I would be a civilization able to use (or store) all the energy available to its home planet.

Type II would be a civilization based on harvesting the radiation of its star, going well beyond our current conception of solar power into various levels of Dyson engineering, as described in the previous section.

Type III would involve capturing the radiation of multiple stars, up to the galactic level.

Kardashev's scale has gained credence because humanity appears to be well along in the process of attaining Type I status, which is assumed to involve a civilization with access to an amount of energy equivalent to the total solar power received by its home planet. For us that's about 1.74E17 watts, or about 10,000 times more energy than humanity currently uses, counting the both the energy on the power grid and the energy involved with off-grid transportation and heating, etc. With a moderate growth rate, we could be there in a couple of centuries.

Getting to Type II is harder to envision with current technology, as it would involve capturing essentially all of the sun's radiation, giving us access to 4.0E26 watts. We'd probably be more likely to get there by developing sources of equivalent power, such as fusion, antimatter, artificial black holes, or other technologies currently unanticipated. Again, there is no basis for predicting what such power would be used for, but it seems safe to say that a use would be found.

Type III is even harder to envision, as it would require interstellar if not inter-galactic travel at speeds convenient to human commerce, plus the near-instantaneous transmission of harvested energy. Dyson-style engineering to capture the luminosity of an entire galaxy (4.0E37 watts) is not conceivable—but would not be necessary. There would be individual power sources whose output, if harvested, would equal the luminosity of a galaxy, such as larger black holes, gamma ray bursts, and quasars.

Kardashev stopped there, but others were inspired to press on, postulating further types, such as Type 0 (based on Victorian technology, in the megawatt range), Type IV (capturing a level of power equivalent to that produced by the visible universe, or about 1.0E45 watts) and Type V (capturing the power of multiple universes, offering perhaps unlimited watts.)

But Kardashev's scale may be seen as something more than a long-range theoretical prediction—it could be seen as a warning. Advancing from one level to another would not happen smoothly, and there would be friction between those who have attained the new level and the laggards who think the old ways are best. For instance, recent history exposed a lot of friction between those who embraced the industrial revolution, and those on the revolution's periphery. In the resulting interactions those who represented the industrial revolution usually enjoyed an advantage—but that advantage was not absolute, as Gen. Custer could attest. But the difference between arrows and firearms is likely to be invisible compared to the difference between Type I and Type II, or Type II and Type III, especially when expressed in combat, as even the laggards will possess mind-bogglingly destructive power. Kardashev was not laying out a path to an idyllic future. Annihilation will remain a threat.

As for detecting extraterrestrial civilizations that might be climbing the Kardashev ladder, SETI has long but unproductively sought broadcasts from the stars that are similar to our own, with its budding Type I civilization. But these searches assume that any sender would have attitudes toward communications that mirror our own. An example of the dangers of such an assumption may be napping at your feet: a house cat with a receptive vocabulary of perhaps 25 human words. It has found ways to trigger your food-bowl-filling behavior, but shows no desire to discuss metaphysics. Meanwhile, parrots and apes have been taught to use spoken words or sign language with their keepers but never thereafter use it with each other. Dolphins on the other hand clearly consult with each other, presumably with some kind of sonar language, while snubbing us.

As for Type II civilizations, they would be evidenced by stars with odd variability, as huge orbiting solar radiation collectors periodically block our view. As mentioned in the previous section, the so-called Tabby's Star is a candidate, and other stars with odd periodic variability are also under investigation.

As for spotting Type III civilizations, a 2015 star survey that sampled infrared light sources in wavelengths associated with waste heat did not find any that were not obvious astronomical objects. The researchers concluded that Type III civilizations must be rare or nonexistent in the visible universe. While that's a valid point, if a Type III civilization was harvesting a vast amount of energy in one locale, the tell-tale might be the absence of waste heat rather than its detectable presence.

As for Type IV and Type V, with galactic and pan-galactic mastery, we would have nothing to compare it to, and would not know what to expect if we were looking at one. One such civilization could be active and generating the universe that we see, and we could only assume that what we are seeing is the result of natural processes.

Meanwhile, keep in mind that Kardashev did not actually lay out any technology blueprints for getting to his levels of energy consumption—he assumed that the necessary technologies exist, and could be discovered and exploited without undue risk of annihilation. Clearly, advanced energy sources will be required—getting 10,000 times more energy on-line than we currently consume is not going to happen with petroleum-based technology.

What's been seen in the sky, of course, sends the same message that Kardashev implied—clearly superior technology exists. It awaits our discovery.

◆   ◆

1.5 - Advanced Energy: "Daemons"

You take an upright barrel and drop a wall into the middle of it, evenly dividing it into two volumes. You pour boiling water into the right side. Then you pour an equal volume of frigid water into the left side. Since we didn't say the dividing wall is a perfect insulator (since it never is) the two volumes of water will eventually take on the same tepid temperature, the hot water cooling and the cold water warming until equilibrium is achieved.

And that is an example of the Second Law of Thermodynamics in action. The law states that any isolated system will spontaneously settle into an eventual state of thermodynamic equilibrium, with an even spread of the available heat.

But let's assume the divider was not passive sheet metal, but was instead something we'll call a daemon membrane. The membrane would test the temperature of individual water molecules and pass the warm ones to the right and the cold ones to the left. So the hot water would never get cold and the cold water would never get hot.

Wouldn't this be a violation of the Second Law?

The general idea was proposed as a thought experiment by pioneering physicist James Clerk Maxwell in 1867, and the conundrum was soon called Maxwell's demon, although he never used the term. (We're using the "daemon" spelling, from the original Greek, as it implies a potentially benevolent entity.) The question is more than Victorian navel-gazing, as can be seen when you add the following refinements to our example.

First, you connect a feed tube from the top of the barrel's right side (i.e., the hot side) to a toy model steam engine. You connect another tube from the exhaust outlet of the toy's piston to the top of the left side (i.e., the cold side) of the barrel. Steam will expand up the feed tube into the toy and do work in terms of pushing its piston back and forth and spinning its little flywheel, before venting into the other side of the barrel to condense. As the Second Law takes hold the steam will stop gushing through the toy and its flywheel will cease spinning as the system settles into an even, tepid state.

But if you replace the divider with our daemon membrane, which segregates the molecules according to their individual temperatures, the hot side of the barrel will stay hot and the cold side will stay cold. The flywheel on our toy will continue spinning indefinitely—it will become a perpetual motion machine, continuing to do work without the input of any energy. Of course, that's anathema to science. Cosmic laws don't have loopholes, but it looked like Maxwell had found one.

The consensus among pundits since then is that there really is no loophole, for two practical (rather than theoretical) reasons.

The first reason is that no daemon barrier exists that could sort, sequester, and transport individual molecules according to their temperature or any other criteria, and none is contemplated. This used to be a conclusive answer, but considering now much intelligence is being crammed into small, slab-like objects these days (played a video game on your cell phone recently?) the answer seems less and less convincing.

The second reason is that if such a daemon membrane did exist it would be an active rather than a passive barrier. It would have to gauge trillions of individual molecules and then do something with them one at a time, and the process would inevitably require the expenditure of energy. The end result would not be a perpetual motion machine, since energy would have to be added from the outside to keep the membrane daemonic and, in the end, to keep that wheel spinning. So the cosmic order (as guarded by the laws of thermodynamics) would not be in danger. The efficiency of the daemon would determine whether it's fated to remain stuck in that barrel with the spinning toy, or to become the foundation of a new generation of power generators. (The daemon could well prove more efficient at raising steam than the old method of applying heat to the outside of the barrel.)

But if we had a daemon membrane (and remember, so far, we don't) why stop at sorting individual molecules according to their temperature? Why not sort them according to their identities? If you're filtering water, you could have it pass only the water molecules and leave everything behind. Suddenly you have a desalination plant, and can make deserts bloom using seawater. (Of course, we already have desalinization technology, so the question would be how efficient the daemon would be compared to existing systems.) With further processing, the water can be broken down into oxygen (about 86 percent by weight) and hydrogen (about 11 percent by weight.)

Maybe you don't want the water? Then the daemon can pass the solids. After you get past the sodium and chlorine (i.e., the salt, about three percent by weight) in seawater, there's also recoverable amounts of magnesium, sulfur, calcium, potassium, bromine, carbon, and vanadium, plus traces of anything else you'd care to name.

If you want to stay dry a daemon ought to work as well with fresh air, getting you nitrogen (about 78 percent by volume), oxygen (21 percent) and argon (one percent), and water in vapor form (usually about 0.25 percent), plus traces of carbon (from carbon dioxide and methane), neon, krypton, and even free hydrogen and helium.

Visitors have been seen simply floating in the air, with no apparent interest in traveling anywhere. Perhaps they're simply mining the air. If they use matter-antimatter or fusion/antimatter fuel (as laid out in Section 1.1 and 4.4) they'd want to extract small amounts of hydrogen, which can be acquired from water vapor. (Others mining water while underwater would remain unseen.)

The daemon, and other forms of processing and refining, would require energy. If the visitors in question have tapped into the energy sources hinted at in Section 2.1, they have that energy—in fact, they may literally have more energy than they know what to do with.

Humanity has been developing microelectronics for several decades now, and chips with circuit geometries in the 5 nanometer (nm) range have reached the market, with 3 nm being the next goal. But we have only recently begun exploring parallel nanotechnology, which would involve the manipulation of matter (or making devices with moving parts) at similar nanometer scales. (A nanometer is one-billionth of a meter, or a millionth of a millimeter. The average bacterium is about a 1,000 nm long. About eight carbon atoms would fit in a nanometer.)

Our experience with microelectronics shows that after you've mastered microscopic transistors you can combined them into logic gates and memory cells, and then scale up to billions of identical components on the same device—and suddenly you're playing video games on the subway. Parallel developments with nanotechnology won't happen overnight, but we may get our own daemons sooner than you'd think.

Assuredly, they'll need their own power sources—the Second Law will remain safe.

❖ ❖ ❖

2
Immaculate Energy: Overview

To accelerate to high speeds without needing huge fuel tanks, while at the same time leaving no heat trace or exhaust, implies something that goes beyond the advanced energy sources discussed in the previous sections. It implies a portable source of immaculate energy, i.e., one that is effectively inexhaustible, but does not pollute, and does not contribute to global warming.

We have no such technology now, and as a result our feckless use of resources poses increasing dangers for our economy, our culture, and even our global environment.

The problem is not the over-consumption of a particular resource. As it turns out, market mechanisms handle availability issues pretty efficiently, upping prices for items that become scarce, thereby motivating consumers to find alternatives. And where there is one technology there are generally alternative technologies: gasoline can be made from coal as well as refined from petroleum, for instance, albeit it at a higher cost.

The problem is that the same mechanisms have no impact on the way we use resources after we've acquired them, and the blithe way we use some of them has led to a buildup of noxious byproducts. The result is like birds fouling their own nests. The burning of mass-produced, cheap fossil fuels since the Industrial Revolution is blamed for an upsurge of greenhouses gases (principally carbon dioxide, methane, nitrous oxide, and fluorocarbons) in the atmosphere. This upsurge is blamed for noticeable global warming and associated climate changes.

A previous generation had dreamed of a modern civilization running on the electricity provided by clean, safe, reliable nuclear power plants. The burning of fossil fuel would become a thing of the past as the populace switched to electric vehicles that took advantage of the cheap electricity made possible by the fact that the cost of fuel for nuclear power plants was economically insignificant. But, again, market mechanisms did not initially address the impact of using the resource. When the cost of processing nuclear waste, the cost of the eventual decommissioning of obsolete plants, and the cost of periodic accidents caught up with them, nuclear plants were left struggling to compete economically with old-fashion generating plants powered by natural gas.

In other words, it was not the immaculate power source they rightfully yearned for.

Basically, nuclear power plants are steam engines that use the heat of nuclear reactions to boil water. A huge amount of heat is generated by heavy equipment to generate a huge amount of energy. A large amount of radiation is also generated, but the very bulk of the equipment provides shielding, as designed.

But what we see in the sky involves craft that are clearly using huge amounts of energy, but leaving no trail of pollution, heat, or radiation. Nor are they tucked atop huge fuel tanks that will be empty in ten minutes. Apparently, their source of energy does not involve the consumption of a resource that generates motion and, as a byproduct, heat. Clearly, what they are doing is drawing energy from the universe itself, taking advantage of the nature of space and time.

In other words, they use immaculate energy.

Such energy is not the province of dragons and wizards. Modern science points to several possible sources, as explained in the next sections. So far, no technology has been proposed that could tap this energy on a large scale, but it appears that if we were able to tap it the laws of thermodynamics would become a dead letter, since the source would be, in practice, inexhaustible. Basically, we would be using the boundless universe to supply power to a human-scale machines, and the universe is not likely to ever notice. The challenge might be to find a safe use for the vast amount of newly available energy, which would likely come with no fuel costs.

Our ignorance should not discourage us. As mentioned in the Introduction, about 1820 there was some question whether electricity—then the stuff of parlor tricks—had any serious use. Now, you're probably reading this under electric lights, if not on a computer screen.

But it's fair to point out that if we were able to master immaculate energy in the near future, that mastery would not free us from tough choices concerning the environment. We would still be altering the terrain by building cities, roads, and farms, and we might be able to do it even more blithely than we're doing it now.

But fossil fuel would be out of the equation, global warming as a byproduct of pollution would be discussed only in history books, and we could pass on the Earth as we inherited it.

❖ ❖ ❖

2.1 – Immaculate Energy: Zero Point Energy

According to quantum physics, every point in space has to be able to service the demands of the Heisenberg Uncertainty Principle, which says that you can't precisely know both the location and momentum of a subatomic particle, thanks to the wave-like nature of such particles. (The more you know about one property, the less you can know about the other, says the principle.)

Consequently, a subatomic particle can never be motionless (even when its temperature is at the coldest possible "zero point") since then you'd know both its speed and momentum (both being zero at the "zero point."). Consequently, there must be some kind of background "zero point energy" (ZPE, we'll call it) that the particles tap to keep up the motion. To that end, there must be a different mode of ZPE for every kind of forcefield that the particle might have to react to. And ZPE must be available anywhere in the universe where you might find a subatomic particle, which means every geometric point in the universe, without exception.

The level of available ZPE is calculated in joules, a basic measure of energy equal to one watt-second. (One AA alkaline battery would store 9,360 joules.) Meanwhile, a widely accepted calculation puts the number of joules of ZPE in a cubic centimeter (a cubic volume about equal to the end of your little finger under the nail) at 1 with 107 zeroes behind it.

Inconveniently, we don't have ready names for numbers that big, although you might try "ten million times a googol." So, we'll render it as an exponent of ten: 1.0E107 joules per cc.

Conceptualizing such a number is no easier than naming it. The amount of electricity consumed by humanity in 2017 was 2.6E15 watt-hours. A watt-hour is equal to 3,600 joules. This means the ZPE in one cubic centimeter could power the Earth for 1.07E94 years, which not only is longer than the visible universe is thought to have existed, but is also longer than it is likely to exist in its current form.

Expressed as BTUs (the unit of heat that raises one pound of water one degree Fahrenheit) there would not only be enough energy in one cubic centimeter to boil Earth's oceans, but to boil them billions of billions of times: 1.8E80 times, or about one oceanic boiling for very subatomic particle in the observable universe. (We're assuming 3.2E8 cubic miles of water weighing 2.94E21 pounds that would need to be raised 177 degrees, since the average ocean water is a bracing 35 degrees Fahrenheit. Meanwhile, a joule equals 0.000948 BTU.)

With numbers like that, energy is no longer a limitation. If you want to do something, and the chief requirement is energy, you can do it. Would the equivalent of Hoover Dam's output be helpful? Would a quadrillion Hoover Dams be better? Then there's no problem. It's all there in your little finger. It's as if the universe were offering us a free lunch, in terms of energy.

Use of the word "infinite" seems tempting but would be misplaced. These numbers are gargantuan, but they're finite and therefore infinitely short of infinite. We are unaccustomed to thinking of numbers with such scales, especially in terms of energy, but if we start doing things like warping space-time for convenient interstellar travel (as opposed to one-way multi-generational migrations, see Sections 6.1 and 6.2) we might get used to it.

Indeed, the behavior of certain visitors, maneuvering in ways that should demand vast amounts of energy, is no longer a mystery, assuming they're tapping ZPE.

Meanwhile, there is no problem with the laws of thermodynamics, since tapping the energy does not drain a heat reservoir. The energy is part of the structure of the universe, and will be replaced as it's extracted.

Of course, things are never so simple, and there are two huge caveats to be considered before planning an energy-soaked, ZPE-powered future for humanity.

The first is that plenty of pundits dismiss the huge numbers as the meaningless result of mathematical recreations, with no connection to the real world. For instance, according to the Theory of Relativity, large concentrations of energy exert gravity just like mass exerts gravity, although the effect is rarely noticed. But with such huge levels of energy suffusing the universe, there should be so much gravity in the universe that it ought to instantly collapse into a geometric point and blink out. Clearly it hasn't. So, it's argued that there must be some factor countering that tendency—but that implies that all the variables necessary to calculate ZPE have not been identified. (Indeed, as we'll discuss in Sections 6.6, 7.0, and the Conclusion, what we call quantum mechanics and relativity may merely be the surface manifestations of underly phenomena that we are not in touch with, meaning that we have most certainly not identified all the necessary variables.)

Either way, the other major caveat is that if ZPE is real (and general agreement is that it is indeed real, disagreements centering on the scale of the available power) we have almost no physical evidence of it, nor any direct way to interact with it. Whatever its value may be, we can't use it unless we can interact with it physically, and so far the results of attempted interaction have been meager.

The only physical evidence we have of ZPE, and the only apparent interaction we have with it, is the Casimir Effect, seen when two uncharged conductive plates are placed nanometers from each other in a vacuum. They should have no interaction, but thanks apparently to ZPE they will attract each other, the attraction increasing as the separation gets smaller. (With a separation of 10 nanometers, the force is about equal to atmospheric pressure, or about 16 pounds per square inch. Victorian steam engines often operated at about 50 pounds per square inch. A nanometer is a billionth of a meter, or a millionth of a millimeter.) While the force is slight, and could be caused by something other than ZPE, the energy measured in the attraction is quite close to the value predicted based on ZPE. Considering that free energy is involved, there has been no lack of effort to put the effect to practical use, including at NASA, plus among the usual perpetual motion fringe.

There have been no breakthroughs. Setting up plates to attract each other has not unlocked the petawatts that ZPE teases us with. But that shouldn't discourage us. Radio waves were described mathematically in 1867, but were widely assumed to be mathematical recreations until they were proven to exist in 1887. Enough people saw the potential value of wireless communications that someone eventually figured out how to dial in.

The value of free, immaculate energy is also obvious. Surely some method will be found to let us dial in. The search will continue.

But, admittedly, there is also the possibility that what we see is what we get, with ZPE. It may be that ZPE expresses itself through the ongoing existence of space, time, and matter, and we cannot consciously extract more energy from it, as we are part of it.

But the evidence in the sky hints at another answer. If an extraterrestrial civilization were extracting ZPE for use in spacecraft, we might expect them to show up just about any place, leaving behind minimal evidence of their comings and goings. And that is what we see. (Admittedly, there are alternate possible sources of energy that they may be using, which will be discussed hereafter.)

If they can extract immaculate ZPE from the universe, then so can we, if not in the next century than in the next millennium. The last 250 years shows that we are pretty good at developing technology, and our society is too competitive, and the advantages of ZPE too compelling, to believe that ZPE would not be relentlessly pursued once a large-scale extraction method is identified. Accepting the evidence of what we see overheard, we can assume that our future will eventually include ZPE, and thereafter any energy crisis will stem from an overwhelming abundance of energy.

◈　◈　◈

2.2 - Immaculate Energy: Vacuum Energy

Vacuum energy is a variant of the zero point energy (ZPE) that was discussed in the previous section. While ZPE assumes the presence of matter (presumably, in fact, matter can't exist without ZPE) vacuum energy exists in the absence of matter. Basically, it is the background energy that otherwise empty space possesses, thanks to the oscillations of ever-present energy fields plus the quantum subatomic fluctuations (particle-antiparticle pairs that erupt into existence and then annihilate each other) taking place there.

Matter can't exist without space (we'll assume, not unreasonably) so vacuum energy could be considered additional values for ZPE, but we are going to consider it separately as there is a lot more space than matter.

Meanwhile, calculations derived from particle physics give a value to the universe's vacuum energy that is 120 orders of magnitude less than the accepted value for ZPE. So if ZPE has 1.0E107 joules per cubic centimeter, the value for vacuum energy is 1.0E-13 joules per cubic centimeter.

In other words, a volume of space the size of the Moon (2.19E16 cc) would yield 2,190 joules. That's less than half the yield of one alkaline AAA battery, which should store 5,071 joules. Expressed as British Thermal Units (the heat needed to raise one pound of water one degree Fahrenheit, at 9.48E-4 BTUs per joule) the Moon's volume should yield almost exactly 2 BTUs, or enough heat to warm an eight-ounce drink by four degrees Fahrenheit, or about as much as you'd expect to warm it by holding it in your hands.

Compared to the value assigned to ZPE, this is nothing. On the other hand, the value assigned to vacuum energy is not swirling with controversy, as is the case with the gargantuan values predicted for ZPE, so it's tempting to place more faith in its availability. A non-zero value for vacuum energy is assumed to be responsible for the observed expansion of the universe—and the observed acceleration of that expansion.

But beyond the behavior of the universe itself, there is (as with

ZPE) no obvious way to directly tap vacuum energy. Meanwhile, the small value assigned to vacuum energy might make it seem too trivial to bother with—but of course, that depends on many considerations.

Consider a spaceship that has been accelerated to one percent the speed of light. Then assume it has some kind of energy field around it that can harvest space's vacuum energy in a circular catchment area a kilometer across (i.e., has radius of 500 meters) as the field sweeps through it. Each second, it could extract about 65 watts, or enough to power an old-fashioned incandescent light bulb, suitable for a table lamp.

But space is big, allowing us to increase the scale of the figures. If the collection field were ten kilometers across you could extract 6,540 watts, which could be enough to run some heavy machinery. If it were 100 kilometers wide you could extract more than 650,000 watts, which could help accelerate the spaceship. Etc.

Things get even more interesting for a spaceship that's been accelerated to 10 percent the speed of light—as, for instance, described in Section 4.3. You'd be collecting ten times as many watts at each level, so a collection field 100 kilometers in diameter would yield 6.5 million watts, equivalent to a small hydroelectric power station.

Obviously, everything would depend on having an efficient method for extracting the energy—and that would involve possessing precise details about the nature and value of the energy. We're still in the dark about basics.

Of course, there was a time when we were in the dark about how electricity works, but from our current vantage point there's no way to say that vacuum energy will provide a path to the stars. Nor is there any clear indication that any extraterrestrial civilization is using it, especially as it exists in the shadow of ZPE.

❖ ❖ ❖

2.3 - Immaculate Energy: Dark Energy

Dark energy is a mathematical construction that astronomers have postulated to account for the fact that a substantial majority of the mass and energy in the universe appears to have gone missing. If dark energy is real, and we could find a way to access it (and we have no clue as to what those methods would look like) it could be another source of immaculate energy, albeit one whose scale is substantially smaller than vacuum energy, as described in the previous section, and especially zero point energy, as described in Section 2.1.

To understand the universe's mass-energy shortfalls and the resulting excitement, keep in mind that an important implication of Newtonian physics is that the farther an orbiting body is from the thing it's orbiting, the slower its orbital velocity will be. The International Space Station is 254 miles above the Earth and has an orbital velocity that's a little under five miles per second. The Moon, which is a little more than a thousand times farther away, has an orbital velocity of 0.635 miles per second. Meanwhile, knowing the orbital velocity and distance of a thing provides a clue about the mass of the body its orbiting.

So when astronomers started getting good looks at distant galaxies they started plotting the orbital speeds of the spiraling arms that many of them have, comparing it to the speeds of the material in the central vortex that many also have. (They could derive their speeds from the red shift of their light—actually clocking their movement would take eons.) They found that the inner and outer material was moving at the same speed. That meant that the outer material was not actually orbiting the galactic centers, but was moving in massive clusters dragged along by each other's gravity. That implies that the clusters contain much more mass than we can see. The missing material is not tied up in huge dust clouds, as we could detect such clouds by the light they block. Nor is it tied up in huge black holes, as they could be detected by their gravitational lensing.

Under the current theory, the shortfall is accounted for by dark matter and dark energy, otherwise unknown entities which we cannot see or detect, but which exert gravity, whose results we can watch. (Under the Theory of Relativity, energy also exerts gravity, so dark energy also contributes to the otherwise unaccounted-for gravity.)

It's calculated that the kind of atoms that make up you, earth, air, fire, water, etc. (called "baryonic matter") account for only 4.9 percent of the mass-energy of the universe. Dark matter accounts for 26.8 percent, and dark energy for 68.3 percent. (Photons, neutrinos, and other oddments account for the rest.)

Meanwhile, studies of distant supernovas indicate that not only is the universe expanding, it is expanding at an accelerating rate. If dark energy is part of the nature of space, then as space expands there is presumably more dark energy, allowing the expansion to feed on itself—as more space is created, the increased amount of dark energy provides a repulsive force that accelerates the expansion.

The energy is uniform across space, rather than clustered, like mass, in stars, nebulae, galaxies, etc. Consequently, while the total may be huge, the amount of dark energy in any one cubic meter of space may be far too weak to detect with laboratory experiments.

In fact, calculations indicate that a volume of space the size of the Moon would yield almost exactly 20 joules of dark energy, meaning you'd need the dark energy contained in the volume of 254 Moons to equal the energy in one alkaline AAA battery. As discussed in the previous section, a fast-moving spaceship with a large, efficient collection field might harvest far more dark energy, but it's going to be about one percent of the vacuum energy that might be harvested that way.

But remember—we don't really know. The fact that we can't interact with it means we can't actually measure it.

And that about sums up what we know (or presume) about dark energy—that name "dark" was not pulled out of a hat. We don't know anything about its nature, and so don't have a clue about how to interact with it, manipulate it or otherwise use it. (The same is true for dark matter, covered in Section 3.4. Meanwhile, we don't know if dark matter and dark energy really have anything to do with each other.) Clearly, either it's a huge factor in the fabric of the universe, or represents a huge hole in our understanding of the universe, or is both.

Indeed, there are those who say dark energy is a scientific version of *deus ex machina*: a plot device introduced in the last scene of a stage play that rescues the main characters from peril and magically causes everything to make sense. But the evidence we see in the sky indicates that there is some kind of omnipresent, unlimited, immaculate energy that we can, with the proper understanding, make use of. And when that day comes the results may indeed resemble *deus ex machina*.

❖ ❖ ❖

2.4 - Immaculate Energy: Your Mind

When you see the Moon in the sky, you're probably open to the idea that its presence there is not related to your act of looking at it—it exists independently of you. It will be in the sky again tomorrow, whether you look for it or not, in a position that astronomers can predict with any desired level of precision, moving in a speed and direction that they can also calculate with arbitrary precision. In fact, they can tell you where it will be a million years from now—and where it was a million years ago, as desired, since the formulae work both ways.

That's the way it is in the macroscopic Newtonian domain of planets, and people, and apples falling from trees. Things happen according to precisely defined laws, which function without our being there to perceive them. If a tree falls in the forest it will generate waves in the air that could be perceived as sound should anyone with functional ears be there to hear it—but those waves will still be generated, regardless.

But underlying the Newtonian domain is the quantum mechanical domain of subatomic particles. The way that domain works indicates that the Newtonian domain (i.e., our world) is, at some deep level, an illusion.

But don't let that fact frighten you. Regardless of how the environment is defined, you're still you and, meanwhile, the illusory nature of this veil of tears may offer opportunities.

As mentioned in Section 2.1, the Heisenberg Uncertainty Principle states that you cannot know both the position and momentum of a subatomic particle. The more precisely you're able to measure its position, the less precisely you can measure its momentum, and vice versa. In fact, this is true of any characteristic of the particle that has dual properties, such as its spin on different axes. The upshot is that knowing the initial conditions of a particle does not let you predict its future condition, unlike the orbit of the Moon or the fall of an apple. Consequently, the position, etc., of subatomic particles is expressed in terms of statistical probabilities, rather than Newtonian certainties. With huge numbers of particles behaving according to probabilities, the results look like matter following Newtonian rules of physics.

The explanation for the uncertainty that was preferred by Werner Heisenberg (1901-1971, 1932 Nobel Prize winner) was that subatomic particles do not actually have mundane characteristics like position and momentum, Or, rather, these characteristics exist only when and as we measure them, and are an artifact of our perception.

This could be taken to mean that subatomic particles exist only when and as we look for them, taking on whatever characteristics we seek at that moment. This is troubling because the Newtonian domain that we think we inhabit is composed of subatomic particles, which we just said don't exist.

Physicists have since established that the perception of a conscious observer (you, your cat, etc.) is not necessary to trigger the measurement effect. Actually, any interaction between a subatomic particle and the Newtonian domain can do it. But the Newtonian domain is a product of the quantum mechanical domain, and member particles of the latter exist only when triggered by an interaction. Apparently, there is some kind of feedback mechanism between the two domains. But the interaction leaves us with a chicken-and-egg dilemma, since one of the domains had to come first, yet they apparently can't exist without each other. Presumably there was some initial triggering act of perception. (At this point feel free to insert your personal religion, or abandon the concept of time and its attendant definition of the word "first," or both.)

The upshot is that the universe can apparently keep clicking along without us, projecting its illusion for purposes known only to itself. The Moon was in the sky before we came along to perceive it, and will likely outlive us. But while we're here we are conscious and can trigger the measurement effect, generating subatomic particles with desired sets of complementary characteristics.

That may not sound like a big feat, but if we could control the perception process with precision we might be able to accomplish something reminiscent of magic: creating matter, such as hydrogen atoms, out of almost nothing. And if we could create hydrogen atoms, we should as easily be able to create anti-matter hydrogen atoms. Brought into contact with hydrogen atoms, they will annihilate each other and provide a source of energy, as described in Section 1.1.

Hydrogen is available everywhere, even in farthest reaches of intergalactic space (where there could be as few as one atom per cubic meter.) Perhaps we could count on scooping it up rather than dragging it along in a fuel tank. Then if we could generate the anti-matter on demand we could avoid the hassle of having to store it in a world composed of matter (with which the antimatter would explode on contact, annihilating itself and an equal mass of its container.)

Admittedly, mentally generating more than microscopic amounts of anti-matter might not be possible. But even microscopic amounts might be enough for a person-sized craft. If not, microscopic amounts could still be enough to power antimatter/fusion engines, where the heat of the matter-antimatter explosion is used to trigger the thermonuclear fusion of the fuel, as described in Section 4.4.

If an extraterrestrial civilization were using this approach, we might see modest-sized space craft, with no apparent fuel tanks, performing maneuvers that demand considerable energy. Of course, that's exactly what witnesses report.

Keep in mind that, no matter how handy this energy source would be, if used as rocket fuel it will never be enough for convenient interstellar travel. As we'll discuss elsewhere, no amount of Newtonian acceleration will let you reach multiples of the speed of light—that will require another approach, as laid out in Sections 6.x.

But once you've reached another solar system and want to make local excursions, it might answer all your needs. Again, this would match what we've seen in the skies.

❖ ❖ ❖

3

Advanced Transmission and Storage: The Problem

hether the energy in question is immaculate or not, it is not likely to be generated at the exact time and place, and in the precise amount, that it's needed. At least that's certainly the case with our technology, and it seems safe to assume that such is also be case with anyone else and their technology. To make up for this mismatch, there will have to be some way to both transmit and store energy. (By transmission, of course, we're referring to wireless methods suitable for spacecraft, rather than the terrestrial power grid that handles most of humanity's power transmission.)

The trouble is that energy generation, transmission, and storage are completely different technologies, facing different limitations—there's no way that transmission and storage are going to be immaculate, for instance. The end result is that transmission and storage look far less promising in terms of development potential.

At least, that's the case with foreseeable extensions of current technology. But storage technology depends on exotic chemistry, and we have just begun to glimpse realms of chemistry far more exotic than anything we have been able to probe in any lab. And our understanding of wireless energy transmission is tied to our understanding of electromagnetism, which may be filled with loopholes.

So hang on—we're in for a wild ride.

❖ ❖ ❖

3.1 – Advanced Transmission and Storage: Wireless Power

In our skies we see visitors in comparatively small craft performing maneuvers that ought to require huge amounts of energy—so where's the fuel?

There may be no mystery. As shown in Section 2.1, a staggering amount of background energy is, in theory, available in the cosmos. The scale of this immaculate energy is easily enough to preclude any need to pack fuel along in visible storage tanks, assuming efficient hardware can be developed to tap that energy. But we don't know what kind of hardware will be necessary to perform that tapping, and in our experience large amounts of steel and copper, etc., are necessary to manage large amounts of energy. So, it is tempting to assume that the visitors are having the energy transmitted to them.

Our own experience is that the transmission of energy is possible. But our experience also shows us that wireless power transmission technology has such pervasive limitations that the visitors are almost certainly not using anything like it. But, on the other hand, our knowledge has such huge gaps in it that we ultimately can't say that they aren't using some sort of transmission—assuming they even need to.

In our use, wireless power transmission falls into the categories of near-field and far-field devices. The near-field devices include contact-free rechargers for electric toothbrushes, biomedical implants, and portable electronics. Effective range is often measured in inches, so such technologies are of no interest when it comes to providing energy for space travel.

Far-field devices are at the other extreme, and are often based on microware or laser technology. Microwave emissions can be more directional than lower-wavelength radio, allowing for concentrated beams. The beams can be converted back to electricity by special antenna-receivers called rectennas. These can be as efficient as 95 percent—although efficiency may be of secondary importance to the visitors, if they truly are tapping vast reservoirs of immaculate energy. The beams, meanwhile, are invisible.

Lasers can be focused tighter, but their beams can be blocked, partially or totally, by clouds, dust, or rain, etc. Their beams may or may not be visible. The photovoltaic power collectors that convert the light into electricity are less than half as efficient as microwave rectennas, but they can be twice as efficient as solar power collectors, since they can be tuned to the laser's wavelength.

(Long before either microwaves or lasers were available, famed but eccentric inventor Nikola Tesla had planned an intercontinental wireless power distribution system with global earth return, and his failure continues to inspire conspiracy theorists. During the so-called current war of 1880-1893 he had defected from DC-advocate Thomas Edison to AC-advocate George Westinghouse, getting rich and helping defeat Edison by inventing the AC induction motor. His planned power broadcasting system would have put Westinghouse out of business in turn. Tesla began construction of his grand central broadcasting tower on Long Island in 1901, but abruptly abandoned the project in 1904. It is now thought that he misjudged the conductive properties of the upper atmosphere based on experiments he conducted in Colorado, and designed a system capable only of draining his bank account and alienating his backers.)

At least on the surface, there are two major reasons that the visitors are unlikely to be relying on beamed power. The first is that such technology, as far as we can tell, requires a direct line-of-sight from the transmitter to the receiver. If a visiting craft is too small for the equipment that taps background energy, then a mother ship large enough for the hardware must always be in sight. That's not the case.

The second is that the transmission of such energy is limited to the speed of light. That they are here at all is a strong indicator (although not absolute proof) that the visitors are not bound by the speed of light, so transmitting energy at that speed would be of limited use. But as we'll discuss in Section 6.6, the whole speed of light limitation may be the exception rather than the rule.

Meanwhile, if they're using multi-dimensional techniques to exceed the speed of light as discussed in Section 6.7, they could be using similar multi-dimensional techniques to transmit power, and there would be no evidence of it to an observer.

That said, science has lately glimpsed vast topics that we know nothing about: dark matter, dark energy, exotic matter, etc. Mastery of any one of these fields might lead to technologies that could eliminate the previous objections.

❖ ❖ ❖

3.2 - Advanced Transmission and Storage: Power-Beam Propulsion

Transmitted energy, as described in the previous section, implies energy that is sent from a generator to a remote receiver, where it is converted to electricity and applied to whatever task the receiver needs performed, such as powering a computer. Beamed propulsion (also called directed energy) implies that the beam of energy is directly harnessed to do work, either by heating something or actually pushing it.

Beamed propulsion offers enough advantages that it is easy to see it as part of the technical repertoire of a spacefaring civilization, for orbital launches and planetary travel. But, as we'll see, it is not going to be the chief reliance of a star-faring civilization.

As for orbital launches, as noted in Section 4.0, even the most powerful chemical rockets fueled by liquid oxygen and liquid hydrogen are not powerful enough to reach orbit without some sort of weight-shedding (i.e., using boosters, or being built in stacked stages.) But direct ascent to orbit, without boosters, would be possible if you just used liquid hydrogen and super-heated it in a rocket engine so that it exploded out the bottom. Section 4.1 describes efforts to supply that necessary heat by building a nuclear reactor into the rocket engine—an arrangement fraught with potential problems.

With a beam-powered rocket you'd do the same thing by beaming energy at the rocket engine to super-heat the fuel. There would be no risk of radioactive contamination, but the precision required to keep the beam on target would be daunting. Directing a single powerful laser beam right up the rocket motor to flash-heat the liquid hydrogen as it enters the motor would not be the answer. The exhaust would block the beam, any imprecision by the beam would probably destroy the rocket, and the rocket has to gradually tilt 90 degrees from the vertical in order to enter orbit, cutting off the laser's line-of-sight into the motor. Presumably, multiple, less powerful beams would be aimed at an exposed heat exchanger on the top of the engine. That way a guidance problem on one beam would not wreck the rocket. There would have to be multiple lasers downrange to maintain contact as the rocket ascended and tilted to orbit. Lasers could also heat the fuel tank itself to push the fuel into the rocket engine, avoiding the need for a pump.

This arrangement basically trades a simple rocket design for a complex ground infrastructure, additionally demanding a municipal-scale power source. Success would also require clear weather.

Using it for launches might be more applicable for places like the lunar surface, where weather is not an issue, there's less gravity to fight, and the necessary electricity could be supplied by solar collectors. The rockets themselves could be crude enough (there'd be no need for fuel pumps, for instance, if the laser also heated the fuel tank) that they could be produced on the moon, aluminum being plentiful in the lunar soil.

As for inter-planetary travel, we could build satellites that collect solar power and beam it at rockets that are already in orbit. But the lasers would be continually moving in their orbits, as would the targeted spacecraft. Multiple lasers and power-generating satellites would be needed. (Beaming lasers from orbit to the ground in order to power rockets that are ascending to orbit is probably not an option due to the unfavorable angle. They would need a line of sight to the engine, which is on the bottom of the rocket.)

Surely it would be easier to skip the power satellites and the orbiting lasers and just build the previously mentioned nuclear-fired rockets. They would offer about the same performance as beam-powered rockets, and building and operating them solely in space would avoid the safety concerns.

What might make better sense would be to retain the power satellites and the orbiting lasers, but drop the rockets. You could use solar sails, as described in Section 1.2, by pointing orbiting lasers at them. (Microwaves would also work.) The straitjacket of the rocket equation (see Section 4.0) falls away, and suddenly the only gating factor is the speed of the photons that are hitting the sails—and they move at the speed of light. Acceleration will be gradual but will be continuous for as long as the beams hit the sail.

Travel between planets will involve orbital transfers, and so a given beam will probably be turned on only when the thrust it provides against the sail will result in a desired orbital correction. When no beams are turned on the sail will be turned sideways to the sun to avoided unwanted orbital perturbations from solar pressure—except when the sunlight is also thrusting in just the right direction. Decelerating into orbit around the destination planet will require a beam emitted from the vicinity of that planet.

As for getting to the stars, beam propulsion offers significant advantages—until you look at the details, which are daunting. With multi-kilometer-wide sails (and a crew equipped to repair them if damaged by interstellar debris) and multi-gigawatt lasers, accelerating up to about 50 percent the speed of light should be possible. Presumably, you could build dedicated power satellites in solar orbit halfway between the Earth and the sun, where they would get four times more solar power than they would get in Earth orbit, but cooling would still be practical. (It would be worth it to position them there because you could get by with collector panels the size of a small city, rather than a small state.) You could also tilt the orbits of the solar satellites so that their lasers have a continuous line of sight to the spacecraft and its destination, so the beam can stay on target almost continuously.

Braking at the destination would be possible by unfurling a smaller sail behind the spacecraft and then unhitching from the main sail, which remains in front of the spacecraft. The beam will bounce off the main sail and hit the new rearward sail, decelerating the spacecraft—assuming humanity has continued powering that beam, aimed at a craft that was sent out perhaps decades earlier by people now long dead. Either way, an excursion module will have to be brought along if anything but the crudest maneuvering is desired. No coordination (like switching the beam off and on as needed for thrust) between the sender and the spacecraft will be possible, thanks to the multi-year time lag for any message.

Philip Lublin, physics professor at the University of California at Santa Barbara, has proposed that beamed propulsion might be best for tiny interstellar fly-by probes that have no intention of decelerating. A wispy disk the size of a dinner plate could still hold significant electronics, and a powerful laser could accelerate it to 30 percent the speed of light in a few minutes, letting it coast the rest of the way.

In the end, beamed propulsion with light-sails appears to provide the fastest option for interstellar travel using Newtonian physics and foreseeable technology, with top speeds even above thermonuclear propulsion (as described in Section 4.2.) But a one-way trip to the nearest star would still take about a decade, and getting to other stars would take multiple decades, and more. Consequently, beamed propulsion cannot be considered remotely adequate for convenient star travel, despite its advantages over the competition.

If visitors were using beamed propulsion to reach Earth we would probably glimpse stray beams or their diffraction patterns. Presumably the visitors would have included in their cargo a selection of smaller craft for local excursions.

But the nimble visitors we see in the sky do not appear to be hobbled by any lack of power for their spacecraft. So while they may be using beamed propulsion at home, it seems safe to assume that they used some other technology to get here.

❖   ❖   ❖

3.3 – Advanced Transmission and Storage: the Storage Problem

You probably never thought of gasoline as a miracle substance. But if it were invented this afternoon pundits would be in awe of its easy handling at room temperature, its ready transportability, its convenient shelf life, its more or less acceptable safety when precautions are taken, and the fact that it gives economic value to that black goo that seeps out of the ground in some locales. But mostly they'd talk about its so-called energy density: 46.4 million joules per kilogram when burned. That's twice the density of bituminous coal and three times more than firewood.

When all is said and done, you can fill the average American family car's fuel tank to its average capacity of 15 gallons of gasoline weighing a total of 90 pounds and (assuming an average fuel efficiency of 26 miles per gallon) drive 390 miles before coasting to a stop (assuming an average speed of 65 miles per hour) after six hours. Of course, you'll have found ample reasons to stop before six hours, and (among other things) refill the tank.

Meanwhile, the rechargeable battery in current use today with the highest energy density is the lithium ion battery, leading to its adoption in the electric car industry. At this writing Tesla Inc. was offering its five-seat Model S electric car with the same range as the average car cited above—390 miles on a single charge. Accomplishing that range (reputedly the highest in the industry) took a battery pack reportedly weighing about 1,380 pounds. That meant the average gas-burning American family car enjoyed an energy density about 15 times better than a high-end electric sedan.

And indeed, getting away from up-market examples, the average energy density superiority of gasoline over lithium ion battery technology, including vastly smaller examples for mobile electronics with densities in the range of 500,000 joules per kilogram, is closer to 100 to one.

Admittedly, lithium ion batteries can be recharged, while gasoline can only be burned once and there is a finite supply of it. The energy used to recharge a lithium ion battery can be collected from renewable sources like hydroelectric dams and windmill farms, and transmitted without further effort over the power grid. Gasoline must be pumped out of the ground as crude oil, delivered to refineries, refined into gasoline, and then delivered to the user, sometimes across oceans in huge tankers that require specially constructed ports. Logic would demand that we drop the use of gasoline and switch to rechargeable batteries, but with that hundred-to-one advantage that gasoline offers simply will not go away. (You will periodically see announcements of new battery technology in a lab, claiming to render its predecessors obsolete, but mass production is another hurdle—don't believe it until you can buy it retail.)

On the whole, it seems that we have been spoiled by advances in microelectronics (with so-called Moore's Law predicting that the power of computer chips will double every few years) and the aerospace field (where we advanced from the first powered flight to a moon landing in 66 years.) Battery power is controlled by tightly defined chemical reactions which are not amenable to radical improvements, plus the efficiency of the manufacturing process used to make the batteries, which is not going to radically improve either. So, closing that 100X gap is not going to happen anytime soon, perhaps never.

On the grid itself, the most common form of energy storage, by far, is "pumped hydro storage," and has nothing to do with batteries. During off-peak hours, when electricity costs less in bulk, water is pumped up-hill into a storage pond or tank. During peak demand times, when electricity costs more, the same water is allowed to run downhill through a turbine generator and produce power at a profit. (In some places they accomplish the same thing using rail cars filled with rocks, rolling up and down slopes.) In terms of classical physics it looks like madness as more energy is consumed than produced, but financially it amounts to a perpetual motion machine.

Clearly, neither of these approaches is applicable to space travel. In the absence any battery technology on the horizon with dramatically (i.e., orders of magnitude) better energy densities, we'd be better off not relying on batteries. And that's what we've done—

the international Space Station, for instance, entirely relies on solar power gathered by large arrays of solar collectors that make it look like a rectilinear dragonfly.

But the ISS also passes through the Earth's shadow once during each orbit and power generation falls to nothing until it emerges. During those periods it runs off a nickel-hydride battery pack, which trickle-charges when sunlight is available.

And that's an example of the way things are likely to be for the foreseeable future: in terms of space or time, power generation will never exactly match consumption, and so some kind of storage will have to take up the slack. But battery technology has not kept up with other technologies, and perhaps has even reached a technological endpoint, like advertising blimps whose helium can only lift so much. We will just have to design around their limitations.

We can only assume that any other technical civilization will be bound by the same laws of chemistry that stymie battery development for us, and they're not relying on them, either. And indeed, one thing witnesses do not report concerning visitors is that their craft suddenly start sputtering and then falling to the ground like stones as their batteries deplete.

What witnesses do report are nimble craft that display no lack of power, leading to the suspicion that they are relying on immaculate power, as described in Section 2.1. If so, at any given moment they literally have more power than they can use, and storage is irrelevant.

That, hopefully, is our future, too.

❖   ❖   ❖

3.4 – Advanced Transmission and Storage: Exotic Matter

According to some versions of the story, when Dutch-Lithuanian-German physicist Daniel Gabriel Fahrenheit (1686–1736) proposed the Fahrenheit thermometer scale in 1724, he set 100 degrees at "about as hot as it gets around here" and 0 degrees at "about as cold as it gets around here." Pundits denounce it as unscientific, especially compared to the Celsius scale, but millions continue to use the Fahrenheit scale for everyday weather measurements. After all, it's designed to gauge their personal comfort.

Likewise, we may have gotten a little too beguiled by conditions in our corner of the universe, accepting them as right and normal just because they allow the chemical processes that run our bodies to operate, while not atomizing us, freezing us, or condensing us into specks smaller than BBs. Elsewhere, wildly different conditions apparently prevail, not only in terms of temperature and pressure but in the nature of the predominate matter, whose behavior we literally can only guess at. The result might be something we can use. Or it might just be as useless as it is exotic, and an eternal reminder that there's no place like home.

A trivial example would be common salt (sodium and chlorine, with one atom of each per molecule.) At high temperatures it can form compounds impossible at room temperature, with varying numbers of sodium or chlorine atoms per molecule. Theoretically, some would have conductive properties that would be useful in electronics—if there were such a thing as high-temperature electronics.

A non-trivial example would be dark matter. As noted in Section 1.3, astronomers invented dark matter to account for the fact that galaxies behaved as if they contained much more matter than was visible. It is now believed that dark matter makes up about 85 percent of the matter in the observed universe, but about the only thing we can say about it is that it does not interact (except gravitationally) with the ordinary matter we're familiar with. (Ordinary matter is called baryonic matter, due to the inclusion of subatomic particles called baryons.) This makes it invisible to direct observation by us and our instruments, both being composed of baryonic matter.

Presumably, dark matter is composed of subatomic particles that remain undiscovered, and perhaps undetectable by baryonic entities like ourselves. But since it exerts and responds to gravity, surely it has conventional mass and we'd therefore feel it if we ran into it, even if we can't see it. To have lasted from the Big Bang to the present implies that it must be stable and that it possesses definite and discoverable chemical and physical properties, some of which would surely be useful. Acquiring and studying even a speck of dark matter could throw open new fields of chemistry and physics.

That hasn't happened, leaving us with indirect evidence and conjectures. Dark matter remains well-named, and some astronomers even argue that a few tweaks to our understanding of Newtonian laws might preclude the need for dark matter.

So dark matter may be our hope for the future, and it may even be the mainstay of the technology used by visitors. Or it may be a mirage.

Aside from dark matter (if it truly exists) there's also a hypothetical counterpart to baryonic matter called mirror matter. It stems from the observation that subatomic particles of baryonic matter interact in a left-handed manner. (Handedness in this case is a math concept concerning the relationship of a particle's direction of momentum and its spin—the particles don't shake hands.) There appears to be no reason for this, leading to the theory that we exist side by side with right-handed matter that we can't see or fully interact with. In fact, we can only interact with it through the weak atomic force (which does not need mirror symmetry) and through gravity. (However, contact would not trigger mutual annihilation, as with antimatter, as noted in Section 1.1) The existence of mirror matter would restore the right-handed/left-handed symmetry that the physicists in the field assume that the universe demands.

But that assumption is about the only grounds we have for believing that mirror matter exists. (It's also called shadow matter or Alice matter.) We can't say there isn't a mirror universe inhabited by mirror-people pursuing mirror-politics, etc., reachable (or not) through some mirror portal (or rabbit hole.) And we can't say there is, either. Perhaps some process during the Big Bang eliminated the possibility of right-handed matter. Or, perhaps there is some distant neighborhood in the universe where mirror matter is the norm. But we can be pretty confident that we are not currently living side by side with a mirror universe, as we would detect the gravity of its stars and planets, etc.

If (as some suggest) mirror matter and dark matter are the same thing, then suddenly we know one additional fact about dark matter: that we can interact with it through the weak force. (We already knew we can interact with it through gravity.)

We can be pretty sure, though, that particles of mirror matter interact with each other the same way the particles of ordinary matter interact with each other. In other words, the laws of physics would be the same in the mirror universe, and using mirror matter would offer no advantage for things like interstellar travel. That doesn't mean that mirror matter couldn't prove useful. Perhaps we could use weak-force tools to manipulate it, starting at a subatomic scale and building from there. If so, perhaps we could use it as insulation (since it doesn't interact with our matter) or as an intermediary for extracting zero point energy (explained in Section 2.1.) Studying it for its own sake would surely prove worthwhile.

And when you're there, say hello to Alice.

(Neither dark matter nor mirror matter should be confused with antimatter, which is discussed in Section 1.1.)

For the last word in rabbit holes, there's tachyons. There are hypothetical particles that always move faster than the speed of light. There is no evidence that they exist, and if they did exist they probably could not interact with baryonic matter. Since particles that travel slower than light, and at the speed of light, are known to exist (and compose the observed universe) the existence of tachyons might assure some balance in the universe, assuming such balance is needed.

If tachyonic interaction with baryonic matter was possible the result might be what's called a tachyonic antitelephone that could be used to make a phone call into the past. Presumably your voice would run in reverse, but could be inverted electronically. Wall Street would love hearing about future stock prices, to the point that any such telephone would probably be suppressed by the authorities. (Confusingly, the term tachyonic is also used in physics to refer to energy fields whose mass equation includes an imaginary number—there are such things in quantum physics.)

At the bottom of the thermometer there's also Bose-Einstein condensates, where visible droplets of matter exhibit behavior usually seen only at the subatomic, quantum level. The potential use for this new phase of matter is still being explored.

Meanwhile, there's stars made of quarks, stars made of super-dense materials, and other matter that's been subject to extreme conditions unobtainable in any lab. And then there's matter inside black holes, where the laws of physics as we know them may not even apply.

Clearly, there's more to matter than the solids, liquids and gases that we're comfortably familiar with. Whether knowledge of exotic matter is the key to unlocking our future, or if it's just an exotic piece of the jigsaw puzzle called science, we now can't say.

❖   ❖   ❖

4

Better Rockets: The Need

Our presence in space, to date, has largely depended on chemical-fueled rockets, called "reaction engines" by purists. Such rockets have such enormous limitations that they cannot possibly be relied on for inter-stellar travel. Whatever methods visitors are using to get here, they are not primarily relying on rockets.

Basically, even with our most powerful rocket fuel, we can't even reach low Earth orbit without resorting to expensive technical tricks—forget about going to the stars.

The limitations stem not from any deficiency in our technology, but from the fundamental physics involved. Consequently, we can be confident that any extraterrestrial civilization would run into the same problem.

The physics in question is Newton's Third Law of Motion: for every action there is an equal and opposite reaction. That may not sound very troublesome—until you're forced to solely rely on it for traveling.

Let's say you took a shortcut home, well after dark, which took you across a corner of the Zero G Rink in the park. You thought it'd be turned off, but it wasn't and now there's no lifeguard on duty to throw you a rope. So you're just floating there, motionless, but would like to get home and feed your unathletic 200-pound frame. You happen to have on you a 16-ounce (i.e., one pound) family-sized can of chunky soup. You throw it away from you at an unathletic speed of 20 miles an hour.

Thanks to Newton's Third Law, you will be nudged in the opposite direction. Since your mass is 200 times more than the mass of the soup can, you will drift off in the opposite direction of the can at about one-200th the speed of the can, or about one-tenth of a mile an hour.

So far so good—but getting out of the rink was rather tedious. So next time you go there you cart along a box with a gross of soup cans (i.e., 144 cans.) When you get stuck you start throwing them one at a time, each at 20 miles per hour, each in the same direction.

The problem is that now, carrying the cans, you initially weigh 144 pounds more than you did last time. Your increased weight initially dampens the nudge you get from throwing each can. True, as you throw more and more cans you will weigh less and less, enhancing how much you're nudged by each throw. Meanwhile, each nudge builds on whatever speed was added by the previous nudge. But the end result is still not enough to make up for the initial burden of all 144 un-thrown cans. By the time you've thrown the whole gross of cans, you'd be cruising at 10.8 miles per hour. Yes, that's about 100 times the speed you reached with one can, but, having worn out your arm, you were hoping to go 144 times faster.

It gets worse: if you bring along two grosses of cans (288) you can reach 17.8 miles an hour. But notice that doubling the number of cans did not double the resulting speed, because your initial weight was higher. Doubling the number of the cans again (to 576) gets you to 27.1 miles per hour, again well short of another doubling. Packing ten grosses of cans will only get you to 42 miles an hour, which is nowhere near ten times the speed you reached with one gross.

In rocket lingo, the soup cans are your reaction mass, which, through the expenditure of some kind of fuel, are propelled in the opposite direction that you want your rocket to go. In your case, the fuel was breakfast, which supplied the energy for your throwing arm. The speed at which the cans were thrown is the exhaust velocity. The nudge each soup can gives you is the thrust. Your weight before throwing the cans is your wet mass. Your weight after the cans are thrown is your dry mass. The ratio between the two weights is your mass ratio. The speed you attain by throwing all the cans is your delta-v (physics lingo for "change of velocity.")

You can calculate your delta-v with any spreadsheet using the 1903 Tsiolkovsky rocket equation, named after Russian mathematician Konstantin Tsiolkovsky (1857-1935) who was one of the first people to formulate it in print. The equation: your delta-v equals your exhaust velocity times the natural log of your mass ratio. The implication of the equation is that, when relying on the Third Law, you're trapped in a downward spiral of diminishing returns brought on by the weight of your unused fuel. For best results you'll want a low-mass fuel that generates the maximum possible exhaust velocity.

Rocket fuel is typically rated by its "specific impulse," which is the number of seconds the exhaust velocity generated by that fuel can support its own weight against one G. The rocket fuel in general use with the highest specific impulse is a combination of liquid hydrogen and liquid oxygen. Brought together so that they explode in the rocket engine, they create a reaction mass of superheated steam with an exhaust velocity of about 2.7 miles per second, which equals a specific impulse of 450 seconds.

Even with that kind of power, it is not possible to build a rocket sturdy enough to endure about 3 Gs of acceleration (typical for a crewed launch) with a mass ratio that can produce a delta-v of about 5 miles per second, which is what it takes to achieve orbit. The answer is to have the rocket shed weight as it ascends, typically by building it in stages—multiple rockets stacked atop each other, in other words. This allows you to discard the weight of the bottom rocket and its empty fuel tanks, engines, pumps, and plumbing, and carry on with the second stage, which starts afresh with its own mass ratio plus an initial velocity, and then the third, etc.

The Saturn V moon rocket, for instance, used three stages to reach orbit. The third stage was later re-fired in orbit to boost the Apollo spacecraft into lunar orbit, and then discarded. Rather than stacked stages, the Space Shuttle used solid fuel boosters mounted in parallel during liftoff and the initial ascent. These were discarded after it reached an altitude of about 28 miles. The Mercury program used the Atlas missile, which discarded two of its three engines two minutes into the flight, while the rest of the rocket continued its ascent with its "sustainer" engine, freed of the weight of the other two engines and their plumbing.

This brings up the side issue of thrust—at liftoff, the total force being exerted by the reaction mass must be enough to counter the weight of the rocket and lift it off the ground. The first stage of the Saturn V, for instance, used refined kerosene with liquid oxygen (with a specific impulse of about 250 seconds) for its fuel instead of more energetic liquid hydrogen and liquid oxygen, since kerosene could readily be pumped into the rocket engines in the volume required to generate the necessary thrust. (Liquid hydrogen is fluffy and hard to pump.) Meanwhile, the thrust must be enough to provide a smooth, efficient multi-G acceleration, while the structure of the rocket must be sturdy enough to withstand that acceleration—hence the previously mentioned practical limit for mass ratios for orbital launch rockets.

But the situation is different in orbit, where anything that can produce an exhaust velocity can generate delta-v. For instance, when the crippled Apollo 13 command module was in its Earth-return orbit with a powered-down computer, its astronauts were told not to jettison their urine, lest doing so push them off course.

Meanwhile, once you are in orbit you are more than half-way to anyplace in the solar system, in terms of delta-v, since you only need about two more miles per second to leave Earth's orbit. With liquid hydrogen and liquid oxygen as fuel, you can get that delta-v with a wet mass that's a little more than twice your dry mass.

But these could be long, tedious journeys, since you'll be using your delta-v to launch yourself into customized orbits around the sun, elongated so that you're brought into the gravitational influence of your destination. There, hopefully you'll have reserved enough fuel to slow down and go into orbit, or land, etc.

Basically, simply blasting over to another planet won't work with conventional rockets. Even if you had 10,000 times more wet mass than dry mass, and used liquid hydrogen and liquid oxygen, your delta-v would still fall short of 25 miles per second, and it would take more than two weeks to get to the Martian orbit at its closest. And, on arrival, you'd need even more fuel to decelerate and land, etc.

Fortunately, there are alternatives to chemical rockets, which we'll discuss in coming sections. They offer many options—especially for vehicles already in orbit. But the most attractive (but currently least understood) alternative would be to drop reliance on the Third Law—which we'll also discuss. (See Section 5.5.)

But there is one huge advantage with chemical rockets: the fuel can be pumped out of the ocean. The hydrogen and oxygen needed for rocket fuel can be acquired by passing electrical current through water (also called H_2O) and breaking it into its components. Using the rockets to loft cargos into orbit, the steam they generate will mostly remain in the atmosphere and be recycled indefinitely.

Meanwhile, the rocket bodies are mostly made of aluminum, which is one of the most common elements in the Earth's crust. Mass-producing them may prove one of the cheapest ways to get to orbit.

So, while chemical rockets have their limitations, they'll probably be around for a long time, especially for orbital launches. But they won't get us to the stars, and no one solely relying on them will ever reach us from the stars.

❖ ❖ ❖

4.1 - Better Rockets: "Skyfall"

As laid out in the previous section, chemical rockets labor under the tyranny of Newton's Third Law of Motion and the Tsiolkovsky rocket equation. With a given engine and fuel producing a reaction mass with a fixed exhaust velocity, the only way to continue accelerating is to burn more and more fuel. But you must bring that fuel along, and its weight drags down your acceleration, trapping you in a cycle of diminishing returns.

The typical response is to broaden the cycle by using light-weight fuel that produces the maximum exhaust velocity: liquid hydrogen burned with liquid oxygen, generating an exhaust velocity of 2.7 miles per second. Even then, you won't be able to reach orbit without using staging or boosters or some similar weight-shedding trick.

It's been proposed that another way to broaden the cycle would be to skip having fuel and just start with the reaction mass—liquid hydrogen, it being the lightest element. You would pump the liquid hydrogen into the rocket engine where it would be flash-heated into a roaring mass that explodes out the motor's vent. You would get the necessary heat by having a small nuclear reactor inside the rocket engine.

To start the engine, you'd use some kind of actuator to bring the fissionable components of the reactor together and "go critical." Just before it starts to melt you hose it down with a stream of liquid hydrogen at -420 F. The hydrogen heats to about 4,900 F on contact and explodes out the bottom. That's about the combustion temperature of the exhaust of a conventional rocket motor using liquid hydrogen and liquid oxygen, but since only hydrogen is present the exhaust weighs less and so reaches about double the velocity of the exhaust of a conventional rocket. The result is a specific impulse between 850 to 1,000 with an exhaust velocity of about 5.2 to 6 miles per second. With that kind of power it ought to be possible to launch a payload to orbit without needing stages or boosters—assuming the engine with its reactor doesn't weigh too much. (Obviously, you'd want these to be reusable rockets, rather than throw away a nuclear reactor with each launch.)

Putting a nuclear reactor inside a rocket engine—what could go wrong?

Well, that's not a happy thought, considering that you're inflicting severe levels of just about every physical stress imaginable on the interior of that rocket motor, including thermal, physical, and radioactive, plus acceleration. It's said that the developers of the Atlas missile (an ICBM later repurposed for space launches) actually boasted that while every single major component of the missile failed catastrophically during testing, no component every failed twice, as they fixed it right the first time. "I'm glad that one got out of the way," astronaut John Glenn reportedly remarked after witnessing the test launching of the same model of Atlas missile that was to later carry him into space, after it exploded one minute into the test.

But philosophical reactions may be harder to come by when launching nuclear reactors if the missile fails and, at best, litters the landscape with radioactive debris. You'll want it to work the first time, not after every component has catastrophically failed once.

Actually, it would appear that nuclear rockets would be an example of technology best deployed and used solely in space. The reactor components could be assembled in space, so that they would not be capable of going critical until then.

And indeed, various space authorities have dabbled with the idea of nuclear rocket and aircraft propulsion going back to the Cold War, when the Americans had a molten salt reactor that seemed to promise simplicity and safety. (The Soviets reportedly handled the problem of heavy anti-radiation shielding by dispensing with it and being philosophical about the resulting casualties.) But the only deployed device at this writing appears to have been the Russian 9M730 Burevestnik (Petrel, or Storm-Bringer Bird), assigned the NATO reporting name "Skyfall." It's a cruise missile, rather than a rocket, meaning it's a robot jet that flies, like all jets, in the atmosphere. It is propelled by a ramjet (after being launched with a booster rocket) that heats the air it ingests using its nuclear reactor and then expels that air to generate thrust. Since it needs to carry neither reaction mass (that being the air it ingests) or fuel (energy instead being provided by the decay of its nuclear material) it supposedly can fly for months at a time.

As for reliability, in August 2019 Russian media reported the deaths of five scientists near Severodvinsk, on the White Sea, 1,200 miles north of Moscow, in a mishap following the failed test of a device that, from the vague description the authorities provided, could have been a Skyfall or an actual rocket using nuclear power. Either way, the victims were apparently trying to recover the device from the seabed when there was an explosion large enough to register on seismographs in Norway, and radiation background readings briefly spiked.

There may come a time when the idea of operating a nuclear reactor inside a rocket engine may be unremarkable. At that point we will have rockets at our disposal with about twice the power of the current ones. But they will still be rockets that have to bring their reaction mass with them. They will still operate under the limitations of the rocket equation laid out in Section 4.0, albeit the straitjacket will be looser. (For instance, generating the 2 miles per second delta-v needed to leave Earth orbit would only require a wet-mass to dry-mass ratio of about 1.5, whereas a conventional rocket would need a ratio a little above 2.) Space travel will still rely on orbital transfers, although we will be able to move heavier cargos and generally enjoy more options.

Star travel would still not be practical, although lofting into orbit whatever infrastructure is required for star travel would be easier. But for actual interstellar travel we'll have to press on for something more exotic.

If any extraterrestrial civilization were using nuclear Skyfall-like rockets, we would not know about it, since, if used safely, from a distance they would be indistinguishable from chemical rockets. And, they would not be powerful enough to enable their users to encounter us, barring their employment in conjunction with some other method for inter-stellar travel.

If used unsafely, we won't be encountering them at all.

❖ ❖ ❖

4.2 - Better Rockets: "Orion"

As explained in Section 4.0, conventional rockets ride atop carefully engineered continuous chemical explosions. The power they can develop is limited principally by the exhaust velocity of the reaction mass produced by those explosions, and the best exhaust velocity we can hope for is about 2.7 miles per second.

There's a brutally simple way to get around that limitation: replace the words "carefully engineered chemical" in the previous paragraph with the word "nuclear." Yes, the rocket would ride atop nuclear explosions, whose exhaust velocity is orders of magnitude faster.

Don't laugh—it's been thought out in detail, and actually could work. Going nuclear should let us cruise the solar system like a yacht cruising the Caribbean. But even knowing that, we still might not want to do it, and the U.S. Government dropped the idea after a 1963 treaty banned atmospheric testing of nuclear weapons.

What was under consideration at the time, "Project Orion" (no relation to the later "Orion" crewed spacecraft project) would have launched a 4,000-ton craft into space by setting off a series of small nuclear explosions under it. The explosions would be the equivalent of tactical nuclear weapons, such as those intended to be fired by field artillery, with a yield of about 0.14 kilotons, or about one percent of the power of the bomb used on Hiroshima. The 300-pound devices would be detonated about one per second under the craft, and about 800 should get it into orbit.

(The source material states that the exhaust velocities of nuclear weapons is classified, but physics indicates it should be between 1,864 and 18,640 miles per second, or between one and ten percent the speed of light. Calculations indicate that the previously described orbital launch would work with an exhaust velocity of an apparently not unreasonable 350 miles per second, assuming the pusher plate catches half the explosion debris.)

For comparison, 4,000 tons was about equal to the weight of the first nuclear submarine, the 320-foot-long steel-hulled USS Nautilus, launched in 1954. It had a crew of 105. The proposed spacecraft would stand upright above a round thickly armored plate, under which the explosions would take place at a precisely calculated distance. The bombs would be shot out of a hole in the middle of the plate that would be re-plugged before each bomb went off. Between the plate and the hull of the spacecraft would be a set of heavy shock absorbers, so the crew would experience acceptable g-forces.

So far, so good—the idea was tested multiple times in 1959, with experimenters boosting models to low altitudes with grenade-sized explosions. It worked. But setting off 800 nukes is probably not something you want to happen in your backyard, or on your continent. The fallout from a single launch might not be lethal (these being small bombs) but the electromagnetic pulse (EMP) would go on for more than 12 minutes (i.e., the time it takes to get to orbit) and would be disastrous for all but the most hardened electronic devices within line-of-sight.

Meanwhile, clearly, if you had an open-ended Orion-based space program with multiple launches, at some point Planet Earth would become one big Chernobyl Exclusion Zone.

And, since you need considerable mass for the shock absorbers and to shield the crew from the gamma rays generated by the blasts, scaling the idea down would not work well, as you would be left with a small, inadequately shielded craft being kicked around like a football. (That would be acceptable, of course, for crewless probes that are built like cannonballs.) Incidentally, while the concept might not scale down, it would scale up smoothly, and there was talk of multi-million-ton Orions, many times the size of the largest aircraft carrier, carrying the equivalent of a small city.

But even if the idea of detonating hundreds of atomic bombs within a few minutes, over a confined region, didn't bother you, you still might not be interested in Orion technology, because those bombs aren't cheap, and they aren't plentiful.

Although non-trivial, cost is probably the lesser of the two problems. The total cost of 800 no-frills tactical bombs, using material from existing stockpiles, would probably be on the order of $250 million. That would be about a two hundred times the cost of the fuel needed to orbit the Space Shuttle, reckoned at $1.4 million. That expense would get you a level of performance 40 times better than the Space Shuttle, with its chemical rockets. (The orbited weight of the Space Shuttle, with a typical payload, was about 100 tons.)

The bigger problem is that atomic bombs can't be pumped out of the ocean. The US spent generations making and refining its stockpile of nuclear bombs, never had more than about 30,000 of all types, and now has about a fifth that number. The Russian numbers are thought to be similar. (The 800 devices needed to reach orbit are, incidentally, more than any second-tier nuclear power possesses.) The devices are made from a hard-to-find, expensively refined uranium isotope, or expensively manufactured plutonium. The most powerful chemical rockets, on the other hand, burn liquid hydrogen and oxygen. Those chemicals are obtained by breaking down water by electrolysis, and of course we have oceans of water. Unlike nuclear material, the hydrogen and oxygen are not destroyed when detonated, but combined back into water (in the form of steam.) So, if used close to home, the fuel is actually recycled.

Many objections go away if you propose operating your Orion-class spacecraft exclusively in space. The fallout would dissipate harmlessly into the boundless void and the EMP would be muted by the vast distances. But space-only operations would surrender Orion's unique advantage of being able to lift Nautilus-sized craft directly from the surface into orbit. An Orion spacecraft would still have to be heavy, and if you're not going to launch it from the surface, you'll have to build it in space after getting the components into orbit. So, space-only Orion technology would be an advantage only if you had cheap access to orbit through some other technology, such as space elevators (see Section 5.4.)

Of course, out there among the stars there may be technical civilizations who aren't bothered about Orion's drawbacks. They may come from biospheres with more radioactivity than ours, so their bodies are hardened to it. And Orion technology could give them access to additional planets in their solar systems where they could mine the radioactive materials that fuel Orion technology. As for EMP, it is possible to harden electronic devices against it—or maybe they don't depend on electronics.

But while they may be using Orion technology locally, they will not be using it for interstellar travel. An Orion spacecraft is still basically a rocket, and like any rocket will have trouble getting beyond ten percent the speed of light, as explained in Section 6.0. If they used some sort of non-Newtonian travel to get an Orion-class craft into our solar system, we would spot it in use, as the rhythmic glint of its blasts would be unmistakable.

If any of those civilizations are using Orions in their home solar systems, light years from this us, we might not spot them, as the glint of the pulsing atomic blasts would probably be lost in the glare of their sun.

❖ ❖ ❖

4.3 – Better Rockets: Nuclear Fusion "Orion"

As stated in the previous chapter concerning "Orion" class spaceships, detonating atomic bombs, instead of liquid chemical propellants, could provide rockets with enormous capabilities. You could launch the equivalent of a nuclear submarine—complete with its steel hull—from the surface of the Earth directly to orbit. (With chemical rockets, by comparison, you're lucky to loft the equivalent of a crowded minivan.) All you'd need to do is set off about 800 field-artillery-scale nuclear bombs in quick succession—and that idea is so scary no one has tried it. (And, as also noted, it would not be exactly inexpensive either.)

But then consider how much more power you'd have—and how much terror you'd invoke—if you used city-killing one-megaton hydrogen bombs rather than bunker-busting atomic bombs intended to fit into artillery shells. The exhaust velocity of these explosions would be several orders of magnitude higher than you'd get from atomic bombs, so your acceleration would be limited not by the available thrust, but by the crew's ability to tolerate g-forces.

We'll assume that the crew has demonstrated a long-term tolerance for one G. (That means they're garden-variety human beings, born and raised on planet Earth, whose inhabitants experience one G at the surface.) Then we'll assume you've built a spaceship weighing about 100,000 tons. (That's the size of a very large, late-model aircraft carrier with a crew of about 4,500.) Of course, it has an appropriate pusher plate and shock absorbers.

By setting off a one-megaton bomb under it every three seconds, you should be able to accelerate it at a rate of one G. (That means you're gaining speed at a rate of 32.2 feet per second per second. The resulting acceleration produces a sensation that is indistinguishable from Earth's gravity.) After ten days of this you'll have detonated nearly 300,000 bombs, and your spaceship will have reached a speed of about 3.3 percent the speed of light. (As in the previous section, the exhaust velocity was not given by the source documents, but the numbers work with an effective exhaust velocity against the pusher plate of 4,500 miles per second.)

From that point you can simply coast, experiencing zero-G, and if you're pointed in the right direction, you'll reach the vicinity of Alpha Centuri (the nearest star system to our solar system) in 133 years.

If the crew is not in suspended animation (and that technology may be more elusive than starship propulsion) you'd need to rig the spaceship as a generational ark, with the crew amounting to a self-sustaining society in a self-sustaining environment. You'd want to design the ship so it could be spun to simulate gravity, which would be better for the long-term health of the crew than zero G. The descendants of the original crew would carry on the mission upon arrival.

If you're impatient, or don't trust kids these days with their crazy music, you can keep up the acceleration for 36 days, and reach 10 percent the speed of light. You'll get there in 44 years, having detonated something over a million hydrogen bombs. (But above 10 percent the speed of light relativistic effects will make further acceleration noticeably harder and harder, as mentioned in Section 6.0.)

There is no mystery involved in designing such a mission— humanity could use its well-established bomb-making skills to reach the stars using thermonuclear Orions, albeit at a snail's pace, cosmically speaking. Designing a generational space ark is entirely possible with today's technology.

Admittedly, there are some drawbacks. The most immediate is that hydrogen bombs don't grow on trees. Detonating one every three seconds, we'd use up every nuclear bomb of any type ever possessed by humanity in about two days. Presumably, more could be built, but producing hundreds of thousands of additional bombs would consume a noticeable fraction of the global economy.

And if gets worse when you consider that the fuel estimate of hundreds of thousands of H-bombs only covers initial acceleration, and not deceleration. Upon arrival at Alpha Centuri, if you decided to stay, you'd need to decelerate in order to go into orbit around it, and that could take about as many bombs as the original acceleration did. Of course, you'd bring them with you—at even greater expense.

Meanwhile, the eventual success of the plan assumes that your descendants will actually be interested in carrying out the original mission, which, to them, may appear pointless, involving far too many inconveniences and dangers, while offering no particular payback. After all, they'll be used to living on the ark. And who's to say that some instantaneous star-travel technology will not have come along while they were in transit, making the whole project irrelevant?

In any event, there is no question that you'd want to operate your thermonuclear Orion solely in space, and not launch it from the surface. Electromagnetic pulse and fall-out would not be a show-stopping issue in space, nor would the cratering of the launchpad and the devastation of the surrounding region. But that leads to a chicken-and-egg dilemma, since to build your Orion-class generational ark in space you'd need to lift heavy loads from the surface, and for that task an Orion is ideal. Short of declaring the biosphere to be expendable, you'd need another heavy lift technology to get cargo into orbit, such as a space elevator (see Section 5.4.)

If an extraterrestrial civilization were operating a thermonuclear Orion around a distant star, we might notice the pulsing flash of the bombs—if we happened to be looking at the right moment. They would only need to operate it for a few days or weeks to get the acceleration they'd need to sail out of their solar system (or decelerate into another one) even if they were shooting for ten percent the speed of light.

If one were decelerating into our solar system, we would most certainly notice, if only from the pulsing static of its EMP. Considerable maneuvering and orbital adjustments would be required before it got anywhere near Earth, so we'd have plenty of time (possibly decades) to prepare a reception, peaceful or otherwise. It seems unlikely that such a craft would be able to land on the Earth's surface, but it might have brought specialized excursion vehicles.

Or the crew just might glance up as their comfortable little world coasted by ours, showing no interest in leaving their accustomed environment to fulfill the will of long-forgotten ancestors. Consider 'Oumuamua, an asteroid that whipped past the sun from interstellar space in 2017 and was then sling-shot by the sun's gravity back into interstellar space. It was roughly tubular and as much as a half-mile long, or big enough to sustain a generational ark. The surface was rocky rather than metallic, but you'd expect it to be coated with debris after a journey that could have lasted eons.

It showed no signs of life.

Maybe next time....

◆ ◆ ◆

4.4 – Better Rockets:
Nuclear/Antimatter Combination

The thermo-nuclear Orion-class spaceship described in Section 4.3 is huge, and, basically, needs to be, because it's intended to use hydrogen bombs as fuel and you can't build miniature H-bombs. In fact, to make an H-bomb you have to start out with a respectable atomic bomb to provide the initial blast through nuclear fission, and then add a mass of fusion fuel (lithium-deuteride) that undergoes fusion in the intense heat of the initial fission blast, generating an even greater explosion. (Also, you want a large ship because you want more than a sheet metal bulkhead between yourself and an H-bomb detonation.)

Meanwhile, as explained in Section 1.1, antimatter would be a premiere source of energy, and could make unsurpassed rocket fuel—if real quantities of it were available. They're not.

But as it turns out, tiny amounts of antimatter—on the scale of the amounts already accidentally generated in particle accelerators—could be bombarded with a few grams of fusion fuel, which would be violently imploded onto the antimatter using ion beams or something similar. The fusion fuel would then erupt into a small thermo-nuclear explosion thanks to the heat and shock of the countless matter-antimatter explosions going on within the fuel's atomic structure. Plans call for a combustion chamber, so that the entire reaction mass is harnessed, instead of setting off explosions behind a pusher plate and harnessing whatever hits the plate. But, like the Orion-class ships, you'd pulse the explosions, at about one per second, and use shock absorbers to cushion the rest of the vehicle.

Conservatively assuming an exhaust velocity of about 82 miles per second, you could generate a delta-v of two miles per second (and thus escape Earth orbit) for a comfortably-equipped craft of 1,000 tons using less than 30 tons of fuel.

That's nothing. With that kind of performance you should often be able to skip tedious orbital-transfers and blast straight to your destination.

But this is another spacecraft that can be operated only in space, if only due to the radiation danger. The crew quarters will require shielding, for instance. (Also, it'd probably destroy the launch pad.) And so, again, we have a chicken-and-egg dilemma, since there are marvelous things we can do in space, but no marvelous way to get there.

The biggest unknown involved in the construction of such a spacecraft would be the manipulation of the antimatter. The tiny amounts required could be supplied with existing or foreseeable methods, but then it would have to be stored for indefinite periods and then, when needed, manipulated precisely. This would have to be done with some form of magnetic suspension. Remember, if you just put it in a jar, the jar would be matter and the antimatter (and a proportional amount of the jar) would be explosively annihilated on contact with each other.

(Of course, another unknown stems from the fact that no one has produced tiny thermonuclear explosions either, but you could have said that about nuclear fission in 1940.)

If there's a candidate for a specific rocket propulsion technology to be the foundation of a space-faring civilization that dominates its solar system, it would likely be this one. The fuel is efficient and (since barely perceptible amounts of antimatter are involved) more or less safe. The quantities of antimatter involved could be produced without draining some fraction of a global economy. The spacecraft can be built on modest, practical scales, and could travel a solar system with the same kind of timeframes that we've come to expect with oceanic cargo deliveries.

If extraterrestrial civilizations are using this technology, there'd be no particular way to spot them from a distance, or even inside our solar system, since the tiny explosions would give off no evident glint unless the exhaust were pointed directly at the observer. They would give off gamma rays, but we are constantly bombarded with gamma rays, in the form of cosmic rays, from various directions.

But efficient as they are, with an exhaust velocity of 82 miles per second, these rockets are still rockets, and are trapped in the cycle of diminishing returns imposed by Newton's Third Law and the Tsiolkovsky rocket equation, since the fuel used to accelerate the rocket also has to be accelerated by the rocket. Even with a ridiculous wet-mass to dry-mass ratio of 10 billion to one, you'd still only get to about one percent of the speed of light and would take more than four centuries to get to the nearest star.

Obviously, something other than rockets is needed.

◆ ◆ ◆

4.5 – Better Rockets: Antimatter Propulsion

In the previous sections we showed how tiny amounts of antimatter, such as we could readily produce, could be used for fusion rockets that could produce convenient travel within the solar system.

But with straight-up matter-antimatter fuel, using equal amounts of both, not only would you have convenient planetary travel, but fuel supplies would cease being an overriding issue.

As mentioned in Section 1.1, combining matter and antimatter would be the ultimate rocket fuel, as the two mutually annihilate each other, producing energy, or energized particles moving at nearly the speed of light. With that kind of exhaust velocity, you should be able to reach orbit with your luxurious 1,000-ton space yacht with a mere 60 pounds of fuel.

Obviously, this is a different world—but don't get intoxicated with the big numbers, because behind the big numbers lurk big problems.

First, 30 pounds of that fuel would have to be antimatter, and we have no way of manufacturing antimatter in quantities large enough to be visible. Some methods, in fact, count individual atoms as they're produced. Meanwhile, all the antimatter would be consumed, with no hope of recycling.

Of course, there is no natural source of refined, engine-ready gasoline flowing out of the ground, yet we have an automobile industry. Perhaps something similar could be done with ordinary matter, and we could find an efficient way to reconstitute it as antimatter. Or, it could turn out that generating antimatter is so unavoidably slow and energy-intensive that there is no rational argument for expending it as fuel—the current situation. But very likely we will find a middle ground: antimatter will be expensive, but practical in some specialized applications.

The second problem is that matter-antimatter annihilation is not perfectly efficient. A lot of highly penetrating gammas rays will be generated, and they will erupt in all directions, not just down the engine exhaust. The crew and electronics will need to be shielded from the radiation, and there'll need to be some serious cooling equipment, eating into the 1,000 tons devoted to your yacht. And then, of course, there's the enormous trouble required to store and manipulate the antimatter without it vaporizing your yacht. For instance, if any intergalactic cosmic ray manages to penetrate into your antimatter fuel tank, you'd better hope it's one of those oddball antimatter cosmic rays, lest it set off a chain reaction as it collides with antimatter atoms and breaks into smaller particles for further collisions, etc.

If all such problems can be controlled, then you could, on paper, have an antimatter rocket engine that's 58 percent efficient. (If the gamma rays could be harnessed, you'd get 77 percent efficiency. But you'll never get complete efficiency due to the relativistic effects of the engine exhaust moving at nearly the speed of light.)

But operating at 58 percent efficiency may seem like no tragedy when such large numbers are in play: it would take about 100 pounds of fuel, rather than 60, to lift your yacht to orbit.

But the drawbacks previously listed, concerning severe supply shortages plus cooling and shielding and momentary accidental annihilation, have soured the pundits, who are projecting the adoption of antimatter propulsion involving something less radical than straight matter-antimatter annihilation.

For instance, in Section 4.1 we described the Russian Skyfall nuclear-powered cruise missile, which injects air, heats it in a nuclear reactor and blasts it out the back, creating thrust. The space-rocket version of the idea would inject liquid hydrogen into a rocket engine heated by a nuclear reactor. The hydrogen would expand explosively in the heat and then blast out the back, generating about twice the performance you'd expect from a conventional rocket. While the operational history of the Skyfall does not inspire confidence in the approach, pundits propose replacing the reactor with an antimatter feed. You'd feed in just enough antimatter to annihilate some of the hydrogen and superheat the rest.

Yes, there'd still be gamma rays—but all radiation would cease as soon as you turned off the antimatter feed. An accident would only endanger the crew and bystanders—radioactive debris would not rain down on the countryside.

Obviously, there are many options for the use of antimatter—on the assumption that, in the future, we'd actually have enough antimatter to play with, and figure out how to handle it safely. And the siren call of antimatter propulsion is such that we won't give up on it easily—it could accelerate us to an appreciable fraction of the speed of light, and is probably the only rocket fuel that could readily do that. For our space yacht, about 200 tons of fuel would get us to 10 percent the speed of light.

But relativity, explained in Section 6.0, means we are unlikely to ever accelerate much above 10 percent the speed of light. For business trips to Alpha Centauri, we'll need something other than rockets and Newton's Third Law.

If extraterrestrial civilizations are using antimatter rockets, we might notice the gamma rays they emit. Interestingly, as previously noted, we are already being constantly bombarded with cosmic rays, some of which might be caused by gamma rays hitting the atmosphere, and we can't really account for the source. Perhaps someday we'll be able to backtrack them—and make come startling discoveries.

❖　❖　❖

5

Non-Rocket Space Travel: Overview

Carefully engineered explosions are not the only way to translate bodily through space. As we'll discuss in this section, we could also use electricity to propel a reaction mass, reaching unheard of exhaust velocities, perhaps drawing the power from the universe itself (in the form of zero point energy) or by bringing along gargantuan quantities in the form of man-made black holes. We could put the vacuum of space to use (since it's not a total vacuum) with jets not dissimilar to the ones operating out of your local airport—just scaled up by perhaps nine orders of magnitude. Reactionless drives that would sort of slide through space, and anti-gravity devices that would let you just float away are not, as it turns out, entirely restricted to science fiction.

And—don't laugh—maybe we could climb a (very long, very strong) rope into space.

While images or fiery rocket launches are closely linked to the idea of space travel, we cannot turn our backs on other methods that have nothing to do with rockets. We have only a few decades of experience with space travel, and have been able to try out only a few methods, tied to a handful of technologies.

It's instructive to look at the history of aeronautics, if only because we've been spoiled by it. In a little over a century aeronautics progressed from a rickety one-person biplane in 1903 to transoceanic airliners carrying as many as 550 passengers. It's tempting to see space technology as an extension of aeronautics (and the wisely abandoned Space Shuttle program seems to have been an outgrowth of that viewpoint) but it's not. A similar pace of development cannot be expected if only due to what economists call barriers to entry. That first biplane could get airborne if it could reach an airspeed of about 30 miles an hour, and taking off into the wind supplied most of that speed. (The Wright brothers soon adopted a drop-weight catapult to get their machine up to speed independent of the wind.) But to get into space, a spacecraft has to accelerate to about five miles a second, and no tricks involving a repurposed windmill derrick, iron weights, and artfully strung cable, block and tackle will suffice. To land, that biplane simply had to glide downward. A spacecraft trying that will burn up like a meteor. Aeronautical pioneers could build their own flying machines by themselves, in repurposed sheds or barns. The initial development of crewed spacecraft required the repurposing of significant portions of the defense industries of two countries, and the Apollo program alone employed, at its peak, about 400,000 people.

But the early efforts at space travel, like the early aeronautical efforts with flying machines, at least proved it could be done. Hopefully, we're at the point with space travel where we can start experimenting and diversifying, possibly using some of the approaches we'll discuss in this section.

◈　◈　◈

5.1 – Non-Rocket Space Travel: Electric Propulsion

At this writing various national militaries are experimenting with so-called railguns, which fire projectiles electromagnetically using massive bursts of electricity. (They're called railguns because the "barrel" of the gun is two parallel rails. The projectile forms an armature between them, accelerating from the bottom of the rails to the top, and then into the air, when power is applied.) Artillery that uses such technology would not be limited to the velocities (approximately 3,000 feet per second) available from detonating various unstable nitrogen-based chemical compounds as a propellant. (Likewise, the gunners would not have to deal with the hazards of accidental detonation that such ammunition poses.) All you need is municipal-level volumes of electricity, but there are naval vessels that can supply that. (The kicker is that you also need carefully engineered components so that the gun won't destroy itself every time it's fired.)

Having read the previous paragraph, and having read Section 4.0 about the practical application of Newton's Third Law, your mind may even now be musing about scaling down the equivalent of a railgun and using it in space. The projectiles could be tiny (gas molecules, even) but you could get an exhaust velocity dozens of times faster than the 2.7 miles per second available with chemical rockets. The thrust may be tiny, but if you keep it going for a while you'll get useable delta-v—and since the volume of fuel used is tiny, keeping it running for long periods is entirely practical. The electricity can be extracted from the environment (i.e., solar power) since we're talking about a few kilowatts, not gigawatts.

You would not be the first to have such musings. About 500 satellites have been lofted possessing some form of electric propulsion, and reportedly about half of modern satellites have it. But they weren't lofted from the surface with the use of electric propulsion—conventional chemical rockets were used for that job. The electric satellites use their electric propulsion for station keeping, orientation control, and orbital adjustment, and a few have used it for orbital transfers. But none have used it to lift off from the ground, since the thrust they generate is tiny.

The use of electronic propulsion goes all the way back to 1962, when a satellite with an ion drive was first displayed. With an ion drive, ionized gas is accelerated using an electric field (the gas becoming the reaction mass, of course.) These were quickly followed by pulse plasma thrusters, which used an electric spark to burn Teflon and generate a hot gas (plasma) that is then expelled using circuitry reminiscent of a tiny railgun.

Hall-effect thrusters showed up in the 1970s, using magnetic fields to expel ionized gas. These were followed by magnetoplasmadynamic thrusters, which likewise electrically expel ionized gas. Since then we've had field-emission electric propulsion (yes, FEEP) thrusters which are like ion thrusters but use a liquid metal (like mercury, etc.) instead of ionized gas. Plus there's been all the technical variants you'd expect. There have also been thrusters that heat and expel liquid, much as the printhead of an inkjet printer. Passing the reaction mass through an electric arc is another technique that's used, similar to an after-burner on a military jet.

These systems typically generate thrust in the fractional or single-digit newton range. By comparison, a baseball pitched at 90 miles per hour by a major league pitcher involves the exertion of 14 newtons. Meanwhile, the Atlas missile used to loft 1.5-ton single-person space capsules into orbit during NASA's Mercury program had a rated thrust of 270,000 newtons. (A newton is acceleration of one kilogram by one meter per second squared.)

A watt can typically produce one newton of power, but satellites typically have a tight wattage budget. As mentioned in Section 1.2, solar power tops out at 1,362 watts per square meter in Earth orbit. Assuming 20 percent conversion efficiency, each square meter gives the satellite about 270 watts, or about enough for four single-bulb reading lamps. Generously assuming 80 percent efficiency by the thruster, by applying an entire square meter of solar power to produce thrust you could get about 215 newtons of thrust, or nearly the equivalent force of 16 pitched baseballs. Applied to a satellite weighing two metric tonnes, that would accelerate it a tenth of a meter per second. That's useful for station keeping, but you're not going to tour the planets that way—the power isn't there, barring acres of solar panels, or your own nuclear reactor.

Or, you could harness zero point energy (ZPE.) As laid out in Section 2.1, the universe theoretically packs enough energy in every geometric point to provide us with a free-lunch, energy-wise. A volume of space equal to the tip of your little finger could easily supply humanity's energy needs (at current levels) from the Big Bang to the present. You'd like 270,000 watts so you can duplicate the thrust of an Atlas missile? ZPE has it covered. You'd like to launch a quadrillion such missiles? No problem. ZPE truly is immaculate energy.

But energy only replaces the fuel of a rocket. The fuel, remember, is the source of the power that pushes the reaction mass out the back of the rocket at high speed. (In a conventional rocket, the burning of the fuel produces the reaction mass.) Even with ZPE as an energy source, you still need to bring along your unused reaction mass, trapping you in a downward cycle of ever-diminishing performance explained in Section 4.0.

And that's why researchers are tinkering (at least on paper) with the quantum vacuum thruster (QVT, also known as the Q-thruster) that would use magnetohydrodynamics (mercifully abbreviated to MHD), to interact with quantum vacuum fluctuations. Reaction mass would be skimmed off from the foam of transitory particles generated by the vacuum fluctuations (to grossly simplify the idea) and would then be given an exhaust velocity using ZPE. Such a spacecraft would need to carry neither fuel nor reaction mass, since both would be harvested from the environment. Its acceleration would be limited only by the g-force tolerance of the crew and by relativistic effects. Zipping around the solar system like George Jetson would be no challenge.

Additionally, moving from star to star with ZPE-powered Q-thrusters would require only the toleration of years of tedium—unless one or more of the theories laid out in Sections 6.x are proven, allowing us to negate or avoid relativistic effects and move faster than light. Huge amounts of energy might be required for faster-than-light travel, but, with ZPE, its covered. Upon arrival in another star system, ZPE-powered Q-thrusters would be handy for local commuting. In fact, the combination is so handy in theory that it's tempting to assume that's what visitors are doing—but at this point we can't be sure which, if any, of the theories in question will turn out to be practical. But of course there was a time when we weren't sure if alternating current, or penicillin, would prove to be practical.

But we don't have to wait for bleeding-edge science to be mainstreamed to make major use of electric propulsion. That railgun barrel? You could make it the size of a soda stray, but two miles long, easily kept stiff in freefall by a light latticework. Let's assume that droplets propelled down it could achieve an exhaust velocity of 200 miles per second. (If the soda straw needs to be longer to get that velocity, there'd be no problem making it longer.) That means your thousand-ton space yacht could produce a delta-v of two miles per second—enough for an orbital transfer—with ten tons of fuel. With chemical fuel it would take about 1,200 tons. You need some electricity, but there are plenty of ways to get that—if necessary, we could slap together something like the power plants of the old Soviet "Whiskey-class" diesel-electric submarines that, during the Cold War, fired their diesels with liquid oxygen.

There's no mystery about getting the electricity, in other words. But such long, spindly spacecraft would have to be built in space, and for that we'd need cheap access to orbit. Again, we face a chicken-and-egg situation, as there are many things we could readily do in space, if we could get there efficiently.

As for our visitors, it seems safe to assume they're using some kind of electric propulsion, if only because there are so many kinds that they could choose from. But we can only guess at the details until we make further progress toward certain scientific horizons. But that will happen—moving toward the ever-receding horizon is what we do.

❖ ❖ ❖

5.2 – Non-Rocket Space Travel: Ram Jets

As mentioned in Section 4.0, rockets use Newton's Third Law to move, expelling a reaction mass behind them, causing the rocket to move in the opposite direction of the expelled reaction mass. The rocket's resulting speed is proportionate to the exhaust velocity and the mass of the reaction mass compared to the mass of the rocket. The reaction mass typically gets its exhaust velocity by being the residue of rocket fuel that is burned explosively in the rocket's engine. That fact that a rocket has to bring along its reaction mass and accelerate the unused reaction mass along with the rocket itself greatly limits what a rocket can accomplish, in terms of acceleration and final velocity.

The jet-propelled airplane that may be roaring overhead as you read this also uses Newton's Third Law, but it doesn't have to bring along its reaction mass. Instead, it uses turbines to suck in the air through which it travels. The ingested air is compressed, fuel is injected into the compressed air, the mixture is made to explode, the resulting hot expanding gas is expelled out the back of the jet engine to become reaction mass, and the Third Law takes over. The airplane only has to bring along fuel.

It'd be wonderful to be able to do the same thing in space—except there's no air there. But there is something called the interstellar medium, which might serve the same purpose. Literally, it's not much to work with, but it might do.

Basically, here on Earth, at sea level, a cubic centimeter of air contains roughly 1.0E19 gas molecules. Inside a laboratory vacuum chamber, the gas pressure (if you can call it pressure) should be about a trillionth of the sea level pressure, but that's still 1.0E10 molecules per cc. Out in interstellar space, the gas "pressure" is about one ten-thousandths that of the vacuum chamber—but that's still a million particles per cc. The particles are predominately hydrogen atoms, with a smattering of helium, dust, and other elements.

That said, the idea suggests itself that if you had a spacecraft with a wide enough scoop (presumably, some kind of force field rather than a sheet metal funnel) and the spacecraft were accelerated to a high enough initial speed by an external booster, it could scoop up hydrogen that could be heated, explosively expanded and then expelled as reaction mass. Yes, you'd have a jet airplane in space—but one whose intake might be the size of a continent.

The arrangement is called a Bussard ram jet, as it was first worked out by American physicist Robert W. Bussard in 1960, and has since been featured in multiple works of science fiction. The incoming hydrogen would be compressed to the point where it undergoes thermonuclear fusion, providing the energy to super-heat the gas and expel it as reaction mass. In other words, it would create a small, ongoing, H-bomb-level explosion, and as long as it maintained a minimum speed would never run out of fuel or reaction mass, since they're both derived from the interstellar medium through which it travels.

The spacecraft's acceleration could go on forever—but it would still never reach the speed of light, since nothing with mass can do that (as explained in Section 6.0, and disputed in Section 6.6 and elsewhere.) But the time dilation stories derived from the theory of relativity could actually be lived out, with astronauts cruising around the galaxy for what seems to them to be 40 years (or some other span less than a lifetime) but on returning to Earth find that eons have passed there. And then they could do it again and again, until they (or their progeny) witness how the universe ends.

Less enticingly, subsequent analyses showed that even while riding its own H-bomb, the spacecraft would be unable to overcome the drag involved in scooping up the interstellar medium at high speed. Basically, the incoming gas would have to be slowed to a stop (relative to the spacecraft) before being fed into the thermonuclear fusion chamber, and the acceleration derived from its exhaust velocity would be less than the energy needed to capture it at high speeds.

The answer would be to not capture the gas, but to heat it as it passes through, like the engine of an actual passenger jet. That would mean that the spacecraft would have to bring along nuclear material whose decay would supply the heat needed to expand the incoming hydrogen. (The concept has obvious parallels to the approach used by the Russian "Skyfall" nuclear-powered cruise missile—when it works as intended, anyway—as laid out in Section 4.1.) Its fuel would no longer be unlimited, although its reaction mass still would be supplied by the environment. It could probably still live out a time dilation story—once, anyway.

But of course no one is going to build a Bussard ram jet to prove to distant generations that relativity works as advertised. Space travel, like any other travel, takes place for defined reasons, and the reason for using a Bussard ram jet would be to get between stars. (Using one for planetary travel would probably not be efficient—by the time you used the necessary boosters to get to the initial speed necessary to ingest sufficient hydrogen, you might as well have used ordinary orbital transfers to get to your destination.) With such a spacecraft you could break the shackles of the rocket equation (as laid out in Section 4.0) and reach a star in years or decades, rather than millennia.

But that it still not sufficient for an interstellar economy. If large enough, such a craft might work for a multi-generational ark ship, but basically it is still a form of rocket (or "reaction engine," of which a rocket is an example) so it is never going to exceed the speed of light.

If another civilization was operating such a craft, it might leave a detectable heat trail, since it's performing the unnatural act of concentrating a swath of interstellar medium, condensing it, heating it, and expelling it. This would be especially evident if one were entering our solar system—but if so, it would be decelerating from its interstellar velocity in order to go into a local orbit. It would soon be going too slowly to ingest sufficient hydrogen to generate any thrust, and would have to switch to auxiliary systems. That would probably make it even easier to detect. That being the case, Bussard ram jets are probably not the main reliance of any visitors we see in our skies.

❖   ❖   ❖

5.3 – Non-Rocket Space Travel: Black Hole Starships

A black hole, of course, is an astronomical body so massive that light (or anything else) cannot escape the pull of its gravity, leaving nothing for a distant observer to see except, well, a black hole. Clearly, a black hole is something a spaceship should have nothing to do with.

Except that (according to the math) a black hole doesn't have to be astronomically massive, just astronomically compact, so there might as well be another universe in there. (More on that later.) Meanwhile, thanks to quantum effects, energy can indeed escape from a black hole, in the form of Hawking radiation (named for British theoretical physicist Stephen Hawking, 1942-2018) and over eons cause it to, basically, evaporate. Smaller black holes would leak radiation faster and have shorter lives. In fact, they could be small enough, and leak enough energy, to serve as power sources.

How small would they have to be? Calculations show that a black hole of about 404,000 metric tonnes (445,333.77 US tons) with a diameter of 1.2 attometers (1.2E-18 meters, or 1.2 quintillionth of a meter) would last about a year and during that time leak about 367 petawatts of power. (The black hole must also possess neither spin nor electrostatic charge, as those factors cause accelerated evaporation.)

In other words, it would have a little more than four times the mass of a late-model fleet-class aircraft carrier, crammed into a sphere 1.2E-18 meters across, making it approximately the size of an electron. During its lifetime it will emit 367 quadrillion watts of power. Divided by the number of hours in a year, that means almost six trillion watt-hours, or approximately the output of Hoover Dam for 1,500 years. In terms of energy density, a black hole of this size is superior to refined gasoline (53.3 v. 46.4 million joules per kilogram.) The energy it emits would mostly be in the form of heat and gamma rays, which should be readily convertible to electricity, for use in some form of electric propulsion. Alternately, it might be possible to put the black hole in a hemisphere so that half the emissions are allowed to escape and generate thrust directly.

Of course, no supplier has such black holes in inventory, but the math shows they could be manufactured, presumably in robotic factories orbiting the Sun well inside the orbit of Mercury, with county-sized solar panels to power massive lasers that would beam energy into a geometric point. The trick would be to concentrate enough power with enough precision to produce an atomic-sized black hole, but again the math makes the production of one per year per robotic factory look possible.

That's the good news. The bad news is that you're giving up convenience in favor of power. Yes, the thing embodies enormous power, but it weighs more than 450,000 tons and can't be subdivided into AA batteries. Therefore, a lot of the power it emits will have to be applied toward moving its own weight. Consequently, the straitjacket of the rocket equation (described in Section 4.0) remains in force, trapping the designer in a cycle of diminishing returns. Used within the solar system, and its relocation accomplished through conventional low-energy orbital transfers, artificial atomic-scale black holes could become a revolutionary power source. Used to power spacecraft—huge ones—they could make the concept of fuel obsolete, but they might also prove more complicated than they could be worth.

Basically, space travel involves huge distances and will therefore unavoidably take significant amounts of time, but atomic-scale black holes will come with strict use-by dates. Meanwhile, the smaller they are, the more energy they emit—but the faster they evaporate.

For instance, the 367-petawatt black hole described earlier would last a year, which is enough for several planetary round trips based on high-energy acceleration, but not enough for even one round-trip using low-energy orbital transfers, and certainly not enough for conventional interstellar travel. Smaller, even more energetic black holes (emitting thousands of petawatts) would last only weeks. Larger (but still atomic-scale) black holes could have lifespans of decades or centuries and could accelerate themselves and a payload to perhaps ten percent the speed of light over the years, and then decelerate as they approached their destination. But they would be far less energetic than our one-year example, producing single-digit petawatts, imposing energy budgets increasingly reminiscent of, well, rockets.

While this approach offers no decisive advantages over any other use of rocket-like "reaction engines" to achieve multi-generational star travel, advocates note that black holes at least pose no mysteries—they're ordinary matter behaving as expected. Antimatter rockets, by comparison, could also produce huge amounts of power, but there is only one fact about antimatter that we are absolutely sure about: tiny mishaps in handling it, that would allow matter and antimatter to mix, would be catastrophic. (See Section 1.1) (On the other hand, artificial black holes would have to be disposed of at the end of their life cycles before they "go nova," but in space there is plenty of room for that.)

Of course, one of the possible multi-dimensional breakthroughs described in Sections 6.x might come into play, so that space travel will not be limited by the need to physically travel. Such methods will probably require huge amounts of energy, which atomic-scale black holes might be well-positioned to supply. Even then, their use-by dates will remain a factor.

Unaddressed by the literature is the question of incorporating a massive black hole the size of an electron into the architecture of a spaceship—it's not like there's flanges on it for bolting it to the deck. Presumably, some sort of force field would be used to keep the spacecraft situated around it, but if the black hole got loose it would pass effortlessly through ordinary matter. (If dropped on the ground, it would free-fall to the center of the Earth with hardly a ripple. On the other hand, it would not have the gravitational attraction to suck up the Earth and destroy it.)

But the simple fact that we are in this universe and we are talking about creating artificial black holes may have cosmic significance—literally. It has been proposed we exist in a multiverse foaming with separate stand-alone universes, including ours; that universes have life cycles and undergo natural selection; and they reproduce through the medium of black holes. The contents of each black hole, meanwhile, would be a universe that resembles the parent universe of that black hole—with random mutations that may or may not enhance the offspring's likelihood to reproduce. Universes with more stars are more likely to reproduce because they are more likely to give rise to intelligent life, with the resulting technological civilizations creating artificial black holes, presumably as energy sources. Our own universe may be the product of the tinkering of some other intelligent race, and by reading this paragraph and accepting its premises you are doing your bit to pay forward the favor and help the multiverse avoid inbreeding and remain healthy and vibrant via the production of new and different black holes. Or you can just walk away....

This currently untestable theory is called the medusa anthropic principle, named from the analogy of the medusa, a gelatinous yet tenacious sea creature with no central nervous system that has an astonishingly complex life cycle. (The medusa is the free-swimming phase of the life cycle of certain jellyfish.)

Regardless of whether we were created by tinkerers in a parent universe, if another civilization in this universe is using atomic-scale black holes for space travel, their creations would emit gamma rays as an engine exhaust and those rays ought to be detectable. However, there are already plenty of gamma rays out there, so the trick would be to detect a moving source of such rays. Even then, the engines would have to be pointed in our general direction. In other cases, it might be possible to build detectors to pick up their activity.

But if we see visitors zipping around the sky, we can be confident they aren't using atomic-scale black holes as their power source—those would have the mass of multiple aircraft carriers, and aircraft carriers don't zip around.

But, of course, they could still be using them for their mother ships, out of view. In fact, perhaps using them is their destiny—or ours.

❖　❖　❖

5.4 - Non-Rocket Space Travel: Space Elevators

From almost every technological approach, cheap access to orbit is the essential first step to becoming a spacefaring civilization. With elaborate, single-use rockets, the rule of thumb is that it costs $10,000 to put a pound into orbit, making spaceflight an expensive stunt done for its own sake, rather than a means to an end.

But why ride a rocket to orbit, when you can just climb there on a long rope? The setup would be reusable and fuel would not be the prime consideration. The idea has been thought out in detail and in theory should work. The only drawback is that building such a space elevator involves a problem that may (or may not) be insurmountable—for now.

Basically, you'd reel down a cable from a spot in geosynchronous orbit over the equator. (An object in geosynchronous orbit will orbit the Earth at the same rate the Earth rotates under it, so the object will remain over the same spot on the surface.) At the bottom you'd probably anchor it to a large ship, so you could maneuver to have the cable dodge space junk. At the top, 22,236 miles overhead, it would be anchored to a counterweight. Tension between gravity and centrifugal force will keep the cable vertical over the anchor point. A machine could then be designed that would climb the cable to the counterweight carrying cargo or passengers, a process that would take about a week. The counterweight at the top would serve as a space station, and spacecraft would dock with it, and depart from it, using orbital transfers, like any other space station—except that outgoing cargo and passengers arrived from the ground without the need for rockets. Incoming traffic likewise debarks at the station and rides the cable to the ground, reaching the ground without the need for dangerous atmospheric reentry.

The cost of putting a pound into orbit becomes the cost of the space elevator divided by the number of pounds it has lofted, so if the system were heavily used the cost per pound ought to be comparatively reasonable, on the order of hundreds rather than thousands of dollars. Multiple climbing machines would presumably ascend and descend simultaneously, spaced to maintain balance and dampen vibrations.

But, as we'll see, getting to the point of having an operating space elevator will take thousands of expensive space launches. Meanwhile, the viability of the design is hard to judge in advance since nothing like it has ever been attempted—but that's not the previously mentioned problem.

That problem involves the concept of specific strength, referring to a substance's strength divided by its density. In textiles, it's usually called the breaking length. Either way, it's the maximum length of a strand or column of the material that could support its own weight when held from the top. For cement, that's a measly 1,426 feet, for low-carbon steel it's about three miles, for aluminum alloy about 13 miles, about 14.3 miles for American eastern white pine, 67.7 miles for spider silk, 159 miles for Kevlar, 2,932 miles for carbon nanotubes, and 3,988 miles with graphene (i.e., lattices of carbon molecules.)

While the space station is 22,236 miles overhead, less specific strength than that will be needed. Fortunately, as the cable extends higher and higher into space the force of Earth's gravity pulls on it less and less, so the cable would actually need to support the equivalent of "only" 3,080 miles of itself. So yes, on paper, with exotic, high-end materials, an adequately strong cable ought to be more or less possible—with some rather large caveats.

The first is that the thing will have to do more than support itself. For the project to be worthwhile vehicles will have to be ascending and descending the cable in a steady stream, presumably carrying standard 40-ton cargo containers. There would need to be a technique that allows ascending and descending climbers to pass each other on the cable. A single climber's round trip to the top and back would take two weeks or more, and if that's all the traffic you had on the cable you may as well go back to building rockets.

The second is that in civil engineering you're supposed to include a safety factor in the design, making the construction stronger than it needs to be, to account for uncertainties, stresses during construction, emergencies during use, etc. The factor varies, from two for ordinary buildings to six for untried material in stressful environments—which surely describes a space elevator. With graphene you'll have a safety factor of only 1.3 for the rope, and adding cargo will cut into that factor.

The third is that you can't pick up the phone and order 23,000 miles of graphene braided into cable. Maybe that's just as well, as it forces us to wait for better materials to come out of the lab.

But assuming that someone did answer the phone and take your credit card number, we'll assume a starter cable about the diameter of a transatlantic data cable, or about 3 inches (or about 7.62 centimeters) in diameter. With graphene weighing slightly more than 2 grams per cubic centimeter, the cable would weigh almost a quarter of a million tons. With rockets that can reach geostationary orbit with 40 tons, it would take more than 6,000 launches, and the 6,000 cable segments would have to be spliced together seamlessly. The empty payload containers would be used to form the counterweight and space station. Once hung, the initial cable would be used to hoist additional cable to reinforce it.

Clearly, such a project makes the construction of the Panama Canal look like a family outing to the beach to build sandcastles. But keep in mind that there was a time when constructing the Panama Canal would have been impossible. But then steam excavators came along, and they figured out now to suppress mosquitoes, and they did construct the canal, transforming oceanic commerce. But it was on the second attempt.

Likewise, the completion of one space elevator would make construction of the next one much easier, as the components of the second could be trucked into space on the first. The third and fourth, etc., will be routine, and we will be a spacefaring civilization. But the construction of the first will be an epochal event.

If an extra-terrestrial civilization were operating space elevators, we almost certainly could not spot it from a distance, but if they have a heavy presence in space (including in our skies) we can presume they had cheap access to orbit. A space elevator would be one way they could have done that.

Of course, if they have mastered immaculate energy (explained in Section 2.1 and elsewhere) they might not need a space elevator (or any similar expedients) to maintain a hefty presence in space. We may reach that point too. But being in space may be a perquisite to mastering immaculate energy, leaving us with the old chicken and egg problem—there are things you can do in space that would be impractical on the ground, but first you have to be in space. A space elevator could end the dilemma. Not that it would then be abandoned—we still use the Panama Canal, mostly for cargo, even as most travelers fly to their destinations.

Keep in mind that other planets could prove much more suitable for a space elevator than ours. Mars, for instance, has only about a third of Earth's gravity, and a geostationary orbit there would be low enough that the elevator could be made out of Kevlar. (Its moons would get in the way, however.) An Earth-sized planet with a faster rotation rate would also have a lower geostationary orbit and so its space elevator would be much easier to build.

So, chances are that someone, somewhere has built one—and we could, too.

◆ ◆ ◆

5.5 - Non-Rocket Space Travel: Reactionless Drives

If we could escape the clutches of Newton's Third Law so that we could act without reaction, the sky would open up to us. We would be able to travel through space without any need to drag reaction mass with us, and therefore without the downward cycle of diminishing returns faced by rockets and other reaction engines. Unavoidably, they need their propellant to accelerate, but must then accelerate their propellant (as laid out in Section 4.0.)

There's been no lack of effort to develop a reactionless drive, but so far the loudest claims of success mostly demonstrated that a vibrating apparatus, placed on a smooth level surface, will manage to move about. (The is called the "slip and stick" phenomenon.) Or, a sufficiently large pendulum atop a wheeled cart will cause the cart to move a little. Repeated explorations of the apparently mysterious powers of gyroscopes to partially defy gravity repeatedly show there is no mystery—inertia adequately explains why a gyroscope wants to point in the same direction even when tilted.

The result has been some science fiction memes, but no working reactionless propulsion systems. Yet, the evidence of what happens in the skies, plus various reports from the far horizon of physics indicates that such a thing cannot be ruled out.

The most widely cited potential source of a reactionless drive is the Woodward Effect, named for California physicist James F. Woodward. Rather than being dismissed as pseudo-science, the effect has been the subject of papers in mainstream journals, but as yet without any clear consensus about whether it works, or should work.

The Woodward Effect is based on the theories of Austrian physicist Ernst Mach (1838–1916, namesake of the Mach number) who decided that we call inertia is not an inherent property of matter. Instead, he decided that what appears to be an object's inertia is instead the result of the gravitational interaction of that object with every other object in the universe, an interaction that is expressed instantaneously. The result is an effect that looks and acts like the property we call inertia. That being the case, there is no reason for inertia to be conserved, meaning that movement ought to be possible without the limitations of Newton's Third Law of Motion (again, see Section 4.0).

Meanwhile, thanks to relativity, the mass of an object changes when energy is added. So by pulsing energy into an object you cause local, transitory changes to its mass, which unevenly alters its relationship to the rest of the universe. Meanwhile, the universe pulls harder on the momentarily more massive parts of the object, and so (under Mach's theories) ought to cause it to move.

Notice that this movement happens without any reaction mass being expelled. The straitjacket of Newton's Third Law drops away. All you need is energy—lots of it, presumably. But according to Section 2.1, the universe appears ready to offer us plenty of that.

An analogy might be children on playground swings who expend energy rhythmically shifting their legs and bodies to change their relation to gravity (i.e., their center of mass) achieving pendulum oscillations that are more and more energetic (i.e., higher)—without touching anything but the swing itself. (The difference is that on the playground the individual swings oscillate but the swing set they're hung from doesn't actually go anywhere, while Woodward's objects are expected to more or less slide through space.)

These theories are not the rantings of a socially isolated crank obsessively tinkering with perpetual motion. Einstein thought that Mach (his contemporary) was on to something. Patents have been issued for Woodward Effect devices, and major research organizations (including NASA) have paid for experiments to explore it.

The results have been inconclusive, some finding the effect loud and clear, some getting mere hints, and some getting washed out.

In 1738, when Swiss mathematician Daniel Bernoulli noticed that air moving at high speed (such as across the top of a flat surface with an added swelling about a third of the way back from the front) exerted less static pressure than air moving at a lower speed (such as across the smooth bottom of the same surface), the phenomenon was hard to measure and of no immediate practical value. Now, we have an aeronautics industry based on it (since it supplies lift to aircraft.) The Woodward Effect may be facing an analogous timeframe, as we learn to measure it reliably and then make use of it. Or, it may prove to be another example of slip-and-stick. In the meantime, no one's laughing at it.

Less widely cited, but also involving less controversial physics, is the so-called Helical Engine proposed by NASA scientist David M. Burns. Trapped ions would be accelerated to nearly the speed of light so that their mass greatly increases. They're then trapped, their impact pushing the craft forward. Applying 165 megawatts of power, it should be able to develop about one newton of thrust. In other words, with more than five times the output of the powerplant of a late-model nuclear submarine, it would develop about the same amount of thrust that your hand exerts against an apple while holding it. (Of course, the thrust would be exerted continuously, and in zero-G the results would be cumulative, and so could eventually generate a useful delta-v. Meanwhile, since no reaction mass needs to be brought along simply to be expended, the spacecraft would be freed from the limitations of the Tsiolkovsky rocket equation.)

But keep in mind that the if we can achieve the promise of immaculate energy laid out in Section 2.1 the energy demands of a Helical Engine would be trivial. And, so far, no one has used the words "slip and stick" to refer to the idea.

Meanwhile, certain reaction-based maneuvers are sometimes confused with reactionless drives. For instance, so-called field propulsion involves the spacecraft taking advantage of an existing magnetic field or other force fields in space to pull itself along. Actually, there is a reaction in such cases, it's just not visible to the naked eye.

Another example would be gravity-assisted space maneuvers, which space probes have been using for decades. Basically, if a spacecraft falls toward a planet but misses it, it can slingshot around the planet and into space, picking up considerable speed in the process. Technically, the speed does not come free of charge, as the orbital speed of the planet that the probe slingshot around will be slowed proportionally—but the change is far too negligible to notice.

The Voyager 2 probe, for instance, launched August 20, 1977, was launched toward Jupiter and, passing it, was able to pick up velocity to go on to Saturn, where it picked up further velocity to slingshot past Uranus and proceed to Neptune—and then leave the Solar System. The trick was careful timing and direction so that each time it missed a planet the spacecraft's new orbit brought it to its next destination.

Power-beam propulsion, as laid out in Section 3.2, is also not reactionless, as the spacecraft is getting a push from the radiation pressure of a laser or microwave transmitter.

Meanwhile, if an extraterrestrial civilization were using reactionless drives, would we know it? Likely, our only evidence would be their presence—and we have that.

❖   ❖   ❖

5.6 – Non-Rocket Space Travel: Anti-Gravity

Gravity (according to relativity) is not a stand-alone force that can be turned on and off, but a distortion in the geometry of space caused by the presence of mass (although large concentrations of energy can also have the same effect.) Meanwhile, sitting on the surface of the Earth, in its gravity well, it is easy to equate weight with mass. When we say something weighs one kilogram, you understand that it would register one kilogram if placed on a scale, and to throw it you'd have to overcome the momentum of one kilogram. If you were floating in zero G it would not register on a scale but you'd still have to overcome the momentum of one kilogram if you threw it.

That momentum you had to overcome is its inertial mass. Meanwhile, as it floats there, it also possesses two additional forms of mass, active gravitational mass (as it attracts other objects) and passive gravitational mass (as it is attracted by other objects.) All three forms are always found to be identical. The actions and reactions of the object can be calculated with precision, with m = I kg.

But what if we changed it to m = -1 kg?

If so, all three forms of mass would have to be negative for that object—and things suddenly get interesting.

Of course, two objects of positive mass (the kind we're used to) attract each other. (In bulk, this force is referred to as gravity.)

Meanwhile, two objects of negative mass would repel each other, all three forms of their masses being negative.

When you have a positive and a negative object, the positive object would use its active gravitation to pull the negative object, while its passive gravitation would be pushed away by the active gravitation of the negative object. The more massive object would win. If you dropped a negative marble out of your pocket, it would fall to the ground, but at a rate slower than a positive marble.

But things get really interesting if you have a positive and negative object of approximately equal mass. The negative object would be attracted to the positive object as the positive object is being pushed away at the same rate by the negative object. If you projected a line through the two objects from the negative one to the positive one to infinity, they would chase each other along that line, accelerating without limit, subject only to relativistic effects. If this looks like impossible energy-free acceleration, keep in mind that the acceleration is an illusion—what you're seeing is a traveling distortion of the space-time continuum.

The use of such an arrangement for spaceflight, is, of course, intriguing. Your spacecraft with you in it would have positive mass, of course, and you'd keep a negative object of approximately equal mass behind it, perhaps in a cage at the end of a scaffold. You'd swing the construction so the apparent thrust would be in the desired direction, and brake by rotating it 180 degrees. To cease all thrust you'd have to keep the construction rotating end over end, or install another equal mass behind the negative one. Orbital launches would not be practical due to the huge mass that would be necessary to overcome the Earth's gravity.

But keep in mind that even with an apparently free source of thrust, relativistic effects would keep the spacecraft from getting anywhere near the speed of light, much less exceeding it. So it's no surprise that visitors have never been reported using the arrangement of masses discussed above.

That doesn't mean, however, that objects of negative mass don't exist or could not be used for interplanetary travel. For instance, rubidium atoms cooled to nearly absolute zero and given reverse spins behave like they have negative mass. Of course, manufacturing negative mass one atom at a time is not going to be practical.

But then there's dark matter. We don't know that it doesn't exhibit negative mass—that's how in the dark we are about it. Since it makes up the bulk of the universe, it would not be surprising if there were quantities of it conveniently located here in our solar system that we could use to make negative-mass spacecraft. But before that could happen we'd have to learn how to detect it and manipulate it.

Alternately, keep in mind that Einstein's famous equation, that energy equals mass times the speed of light squared, works when you plug in negative values, giving us a glimpse of vistas that we are not currently equipped to explore. With a sufficiently large source of energy, such as zero point energy as described in Section 2.1, who knows what we'll be able to do—after we learn not to turn ourselves into soot with it.

Meanwhile, as mentioned, concentrations of energy, in sufficient quantities, produce gravitational effects the same way that mass does, and we might be able to produce the effect of negative mass by that route, perhaps using some sort of negative energy. (In fact, see Section 6.1). Dark energy, which actually predominates in the universe, may amount to a form of negative energy, but we can't really say since we know nothing about it. Using negative energy, rather than negative matter, to produce anti-gravity effects would mean that you are not bound by the size of the negative-mass object you've harnessed, perhaps permitting craft of the sizes reported by witnesses.

The ability to hover without an apparent downdraft is no longer a mystery: if they can generate large amounts of negative energy, they can produce the effect of anti-gravity at will.

Pointing in a slightly different direction is the work of French astrophysicist Claude Poher, who believes that gravity is caused by a flux of massless particles traveling at the speed of light, called universons. These particles could be used explain various astronomical observations without the need for dark matter (see Section 3.4) and are the basis of an experimental reactionless "electric space propelling system" he says he has tested. The device's use of superconductors is reminiscent of the previous, controversial work of Finnish-schooled Russian engineer Eugene Podkletnov, in which superconductor discharges appeared to affect gravity.

In sum, anti-gravity makes a great science fiction plot device, and it would be tempting to dismiss it just on those grounds. But, apparently, an anti-gravity drive is not impossible. For that we have the evidence of mathematics, and of what's been seen in the skies.

❖ ❖ ❖

6
Faster-Than-Light Travel: The Problem

The closest star, the unprepossessing red dwarf Proxima Centauri in the Alpha Centauri system, is 4.244 light years away. (That's nearly 25 trillion miles, or about 40 trillion kilometers.) Let's say you've mastered one of the technologies hinted at in this book that allow a spacecraft to accelerate unrestrained by the need for fuel or reaction mass. Perhaps you're harnessing nearly unlimited zero point energy (ZPE, see Section 2.1) and harvesting reaction mass by skimming transitory particles generated by vacuum fluctuations (see Section 5.1.) Or perhaps you're transmitting fuel to the rocket (see Section 6.4.) That said, you could expect to fire up your spacecraft, point it at Proxima Centauri, and start accelerating at a comfortable 1G (or 9.8 meters per second squared, equal to the pull of the gravity you're accustomed to on Earth.) Halfway there, at the 2.122 lightyear mark, on about the 740th day, you'd being going 627,000 kilometers per second. At that point you'd reverse your thrust so that, on arrival, you'd have zero velocity. The trip would take a little over four years. That may sound like a long time, but remember—Magellan's fleet took three years to make the first trip around the globe, in 1522.

Except it won't work—the maximum midway velocity, as described, is more than twice the speed of light, and you can't go faster than light. In fact, you're lucky to get anywhere near the speed of light. Your simple (if lengthy) jaunt to Proxima Centauri is simply not possible.

It turns out that high-speed space travel is dominated by physical laws whose functioning can be described with mathematical precision, and have been experimentally verified to any reasonable level of proof, but which are counter-intuitive to garden-variety human beings. In fact, they seemed cleverly designed for the explicit purpose of preventing us from reaching the stars.

But, of course, laws have loopholes. But to understand the loopholes, you have to understand the laws.

The foundation of those laws is a physical constant, called c, which designates the speed at which light propagates in a vacuum: 299,792,458 meters per second; or about 1.08 billion kilometers per hour: or 186,000 miles per second; or about 671 million miles per hour.

But with light, the very concept of speed is counter-intuitive. Basically, c is not a speed but an invariant property: light has the same speed relative to you no matter where you are, how fast you're going, or in what direction you're going. If you accelerate to a speed that is a significant fraction of c, your sense of time will slow down and your size will be compressed so that c remains invariant. Your mass will also increase as energy is invested in your acceleration. (You won't notice any of these changes, but distant observers who don't share your acceleration will be able to notice.) Further acceleration will cost more and more energy and you will grow more and more massive—with the cost approaching infinity as you approach c. Infinite energy not being available, you will never reach c if you possess mass. (Even the huge sums of immaculate energy promised by ZPE calculations will fall short—large as the numbers are, they are still finite numbers, and therefore infinitely short of infinity.)

On the other hand, c is the speed at which massless particles (principally photons) travel. Light itself can travel slower than c when passing through transparent media (such as air, water, glass, diamonds, etc.) but reverts to c (without requiring any source of energy to do so) after emerging from the media.

If there is anything that travels faster than c, with or without mass, we don't know about it, and would probably have trouble detecting it directly.

Other implications of the foundational nature of c is the equivalence of energy and mass, as with Einstein's famous equation $E=mc^2$. But since we're traveling, we're more interested in the less famous Lorentz equation, which spells out the relativistic effects of traveling at speeds approaching c.

Treating space and time as a unified structure called spacetime, the Lorentz factor (also called gamma, and named for Dutch physicist and 1902 Nobel Prize winner Hendrik Lorentz, 1853-1928) is the degree of time dilation, length contraction, and mass expansion that an object will experience as its speed approaches c. Time dilation is how much slower its clock will run. Length contraction is how much the object will be shortened in the direction of travel. Mass expansion indicates how much heavier it will be. (Of course, there are other properties, such as momentum, that can be derived.) The formula is one minus the velocity of the object squared divided by c squared, all raised to the power of negative .5.

At 10 percent of c the factor is half of one percent. That may not sound like much but it's enough to throw off rocketry mission calculations and so is usually seen as the speed limit for conventional rockets. (The fact that the fuel weighs more does not make up for the rocket weighing more, since the exhaust velocity generated by the fuel is the same.)

By 25 percent of c the factor is even more pronounced, at 3.28 percent, while at 33 percent the gamma factor is almost 6 percent. After that the factor rises more and more steeply. At 50 percent c the factor is 15 percent. At 75 percent it's 51 percent. At 99 percent it's a factor of seven. At 99.99 percent it's a factor of 71. At 99.9999999 percent it's a factor of 22,361.

At 299,792,457.999 meters per second (i.e., one millimeter per second short of c) the factor is 387,169. With that factor, for every second that transpires in the spacecraft, 4.48 days would transpire for stationary observers back on Earth.

You never get to 100 percent of c if only because, under the formula, that would be division by zero, and under the same rules you can never exceed c.

Obviously, spacetime pushes back against efforts to get to c, implying that space must be a dynamic system rather than a yawning expanse of inert nothingness. Apparently, we have been spoiled by our collective experience in aeronautics, where you can indeed take your craft out and let it rip, blasting through the passive air (so far as allowed by aerodynamic forces and the weather, of course) to your heart's content. But any assumption that space would offer a similar experience, with some added zeroes when it comes to velocities and distances and travel times, must be put aside. It's as if Mother Spacetime doesn't like too-fast movements or too-high acceleration and has laws to prevent such unseemly behavior. The result is not pretty—not only can we not go faster than light, even getting halfway to the speed of light may be too much for our over-taxed technology. So getting to the stars will take not years, but decades and generations, and meaningful interstellar exchanges will not be practical—under the laws of c.

So the race is on to find loopholes in those laws. And the evidence is that loopholes do exist. In fact, as we'll see, there may be more holes than laws, and some of those laws actually contradict each other.

❖ ❖ ❖

6.1 - Faster-Than-Light Travel: Alcubierre Warp Drive

In the previous chapter was examined how spacetime pushes back, making it impossible for anything with mass to exceed c (the invariant speed of light, or 299,792,458 meters per second.) In fact, it pushes back harder as you go faster, and so no spacecraft can ever reach c, much less exceed it. It's as if spacetime has a specific tolerance for the rate of change, and that tolerance is expressed by c. The implication is that convenient and practical interstellar travel, at human time scales, can never happen.

But what if you fixed spacetime? If you could expand spacetime behind your spacecraft and contract spacetime in front of it, you should be able to surf the resulting warp bubble in spacetime to your destination. Better yet, for the spaceship and its crew inside the bubble, no travel is taking place. When seen from the outside, the bubble is clearly traveling, but inside the bubble's "local frame of reference," the crew may as well still be at home. So, according to the relativity and the laws of c, the spaceship and its crew are not accelerating, and so the crippling relativistic effects laid out in the previous section no longer apply. With this simple trick you can indeed travel faster than light—perhaps many times faster—without breaking any laws of physics.

Of course, it's easier said than done.

Mexican theoretical physicist Miguel Alcubierre Moya came up with the idea (now called the Alcubierre drive after his patronymic) in 1994. He said he borrowed the term "warp drive" from the Star Trek TV and movie franchise, which originated in 1966. But the mechanics of the drive were inspired by the calculations involving the inflationary phase of the Big Bang, when objects had to have been moving away from each other at a rate much greater than the speed of light. But the current laws of physics were already in force, so movement faster than c was not possible. Apparently, such movement was possible only because spacetime itself was expanding, and if space underwent warping at that time it ought to be possible to warp it again, on demand. So with a customized, local expansion of spacetime, travel at arbitrarily large speeds—

many times the speed of light—ought to be possible, and it would happen without time dilation or other relativistic effects. In fact, that spacecraft would not experience any acceleration.

Unfortunately, the first pass at calculating the requirements for such a spacecraft showed that the concentrated mass needed to sufficiently warp space was perhaps ten orders of magnitude greater than the mass of the visible universe. Forget it, in other words.

But the idea continued to attract researchers, until observers began speaking of a new "warp drive theory" field of physics. (Perhaps not coincidentally, more "Star Trek" movies and TV spinoffs kept coming out, all sporting starships using warp drives.) After some tinkering, it was decided that you needed only "a few solar masses" to do the warping.

Despite that fact that a "solar mass" means a concentration equal to the mass of our sun, which is 330,000 times more massive than the Earth, the new calculations radically reduce the scale of the requirements—perhaps enough to make an Alcubierre starship physically possible.

Of course, keeping several solar masses on hand to assist with space travel is not a practical solution, and those solar masses certainly can't be turned on and off as needed. But remember how $E=mc^2$? Concentrations of energy are equivalent to concentrations of mass.

Meanwhile, in Section 2.1 we laid out how zero point energy (ZPE), the immaculate power source by which empty space apparently supports the existence of matter, offers the potential of astronomical quantities of energy. Calculations indicate that each cubic centimeter of space harbors 1.0E107 joules. (That's a one with 107 zeroes after it. A joule is one watt-second of power, or the energy needed to accelerate one kilogram one meter per second.) In other words, the cubic volume of the end of your little finger has enough energy to boil the world's oceans, over and over, until you get bored with the stunt. With ZPE, energy is not a scarce resource.

The revised calculation shows that to successfully warp space an Alcubierre starship would need 4.9E30 kilograms of positive mass, and 1.4E30 kilograms of negative mass. By comparison, a solar mass is a sliver under 2.0E30 kilograms.

But if you convert this requirement to energy using $E=mc^2$, you need 4.4E47 joules of positive energy and 1.26E47 joules of negative energy. Together, this is less than half the ZPE calculated to be packed in one cubic centimeter. (And if more energy were needed, remember, there's the neighboring cubic centimeter, and then the one after it, etc.) And, being energy, it can be turned on and off.

So the math works. For all we know, an ersatz Capt. Kirk could someday be commanding a starship blasting through the galaxy at convenient speeds. Except that, while the math works, some of the technology remains unimaginable.

The biggest problem is that, so far (as laid out in Section 2.1) no way has been found to actually harness ZPE. Pundits agree that it does exist, but they don't all agree that the scale of ZPE is as astronomical as is widely calculated, or that we could extract all of it if it is. But it does appear to be truly immaculate—if we drain one cc, it will recharge itself from the rest of the universe.

The second problem is that the warping in front of the spacecraft has to be done with negative energy. Combining the words "negative" and "energy" might sound like an oxymoron, but the math does allow it to exist. However, current technology does not.

Ironically, the original "Star Trek" was set in the 23rd century, or two centuries from now. Given the advances and effort likely needed to perfect a warp drive (assuming one is ultimately possible) that might be about the right time scale.

Meanwhile, if some other civilizations were operating Alcubierre-style starships, they probably would not be operating them in our vicinity. The previously mentioned positive mass needed for warping space, minus the negative mass also needed for warping space, amounts to about one and a half solar masses. If powered-up while parked anywhere in or near our Solar System, the gravitational distortions triggered by such the starship's mass-energy concentration could wreck the solar system, causing the planets to veer out of their orbits, possibly into interstellar space. It could even alter the Sun's path through interstellar space. So such a craft could not be safely operated at full power near an inhabited star system. (The safety margin would probably be smaller when moving faster than light.) Smaller craft, presumably using reaction engines of some kind, would be needed for local excursions. So visitors may be traveling between the stars in them, but we would probably not see Alcubierre-style starships in our skies.

❖   ❖   ❖

6.2 – Faster-Than-Light Travel: String Theory Warp Drive

In the previous section we looked at how faster-than-light travel could be possible by warping space-time with enormous concentrations of mass or energy. Travelers would surf to their destinations in spacetime bubbles at many times the speed of light, but inside the bubbles they would suffer no relativistic effects, as they would be, relativisticly speaking, at rest. But the process would involve speculative science, and would require concentrations of mass (or equivalent energy) equal to several solar masses.

In 2008 two Baylor University scientists (Gerald Cleaver and Richard Obousy) proposed a similar approach to interstellar travel that would involve a warp drive using considerably smaller mass/ energy concentrations, although the science behind it is equally speculative. The end result, however, would be a warp drive reminiscent of the Alcubierre drive described in the previous section.

Their approach is based on string theory, which assumes the existence of several extra dimensions in addition to the three spatial dimensions that you're familiar with (i.e., height, width, and length) (plus time.) The existence of these dimensions is sometimes cited as the reason why gravity is by far the weakest of the four fundamental physical forces (the others being electromagnetism, the strong nuclear force, and the weak nuclear force.) The reason is that gravity is apparently free to propagate through all the dimensions, thus diluting its force, whereas the others forces only affect the three old-fashion spatial dimensions where they reside.

The extra dimensions are assumed to be tiny strings that have circular cross-sections, like garden hoses. Cleaver and Obousy calculated that by changing the size of an extra dimension (by pumping energy into or out of it) they would also change the size of the adjacent spatial dimensions.

In fact, shrinking an extra dimension ought to inflate the spatial dimensions, and inflating an extra dimension ought to shrink the spatial dimensions. Doing this in front of and behind a spacecraft should let it surf through space. Since the spacecraft is not moving even as spacetime warps around it, relativistic effects do not apply, and the vehicle is free to move faster than the speed of light.

The authors based their calculations on a putative spacecraft of 1,000 cubic meters. The energy necessary to do the warping around it, they calculated, would amount to 1.0E45 joules, or within an order of magnitude of the energy you'd get by rendering the planet Jupiter into energy, according to $E=mc^2$.

While that sounds at first like an absurd energy requirement, let's remember zero point energy (ZPE), the immaculate energy discussed in Section 2.1. If it could be harassed, the amount in one cubic centimeter of space (1.0E107 joules, a joule equaling a watt-second) would handily supply the energy needed to move their thousand-cubic-meter spacecraft. In fact, it would satisfy the amount required, squared. (Cleaver and Obousy also indicated a belief that ZPE would be available in the extra dimensions, further magnifying the volume of energy that might be available.)

A string theory warp drive (as we'll call it) also ought to be somewhat safer than an Alcubierre warp drive (as discussed in the previous section.) An Alcubierre drive, when powered-up, would propagate a gravitational distortional in space-time equal to a couple of solar masses. Consequently, going to full power with one inside the Solar System would cause catastrophic orbital distortions for every celestial object orbiting the Sun, including the Earth. In fact, whole planets and even the Sun itself might be drawn into the spacecraft, or try to orbit it. By contrast, a string-theory warp drive would propagate a gravitational distortion approximately that of Jupiter, so powering it up inside the Solar System would still cause dangerous orbital perturbations, perhaps even enough to make the Earth uninhabitable, but would not immediately destroy the Solar System.

Basically, a string-theory warp drive could probably be safely operated closer to an inhabited system than an Alcubierre drive, and the danger may be minimal when it is exceeding the speed of light.

Meanwhile, Cleaver and Obousy calculated that the maximum velocity that could be achieved by a string-theory warp drive would depend on how small you could shrink a dimension. If you could get it down to a Planck length (i.e., the smallest scale at which the laws of physics are still assumed to be meaningful, or 1.6E-35 meter) it should be able to go 1.0E33 times the speed of light. Even a fraction of that velocity would allow convenient intergalactic commuting, although at the high end the energy requirements might be impractical, even with ZPE. Of course, the practicalities of interstellar navigation might often mandate much lower speeds anyway—there's a lot of stuff out there to run into.

Meanwhile, you could not operate a string theory warp drive inside Earth's atmosphere, as it would destroy the planet with its Jupiter-scale mass-energy concentration, Jupiter being more than 300 times more massive than Earth. Operating a warp drive for interstellar travel, you'd need to stand off from your destination and complete the journey with excursion craft using less energetic technology. So what we are seeing in our skies are probably not warp drive starships (unless they can operate at extremely low power) even if they got to our celestial neighborhood using warp drives.

But keep in mind that, so far, we have not managed to find a way to harness ZPE, or get access to and change the size of alternate dimensions, and there are pundits that denounce the entirety of string theory as no more than a source of mathematical recreation with no predictive success in the real world.

But the big picture is that, with this section and the previous one, we have two examples of researchers, using completely different approaches, coming up with versions of warp drives that appear mathematically possible. So perhaps there's something to it, and to interstellar travel.

❖   ❖   ❖

6.3 – Faster-Than-Light Travel: Traversable Wormholes

A wormhole is a theoretical fracture in (or folding of) spacetime that connects arbitrary points in space (or time, or both). It can be thought of as two connected funnels, and anything that enters one side will pop out the other regardless of the distance (or time interval, or both) between the two sides. They could be billions of lightyears apart, or adjacent, in different historical eras, or even in different universes, and the results would be the same—immediate relocation (although in some situations there may be some transit time between the two ends.) Although the end results would be faster-than-light travel, the travelers would experience no relativistic effects.

Spacetime wormholes have been compared to actual holes made in a piece of paper by actual worms. If you draw a line from one such hole in the paper to another and measure the line as having a length of X, you can say that X is the distance between the two holes. If you fold the paper so that the holes overlap, there is suddenly no distance between the two holes, even though the line with the length of X still exists. Spacetime wormholes expand the concept from two-dimensional paper to multiple dimensions (plus time.)

Their potential usefulness in interstellar travel is obvious—if you had one end of a wormhole on Earth and the other on the large terrestrial planet known to be orbiting Bernard's Star six lightyears away, it would be as if they were adjacent. Getting there would be immediate, and not require a lifetime of sub-lightspeed travel. (But the gravity there is probably three time that of Earth's, so you won't want to tarry there.)

But note the use of the word "theoretical" in the first sentence of this section. Wormholes are allowed by the mathematics of relativity, and theoretical physicists have generated reams of formulae outlining their expected characteristics. But so far no wormholes have been spotted, by astronomers or anyone else. If any were left over from the Big Bang (and there may be plenty, for all we know) they'd likely connect random points in the universe. But it's a big universe, with a lot of random points in it, so any random wormhole that we identify is unlikely to be of any practical interest to us.

A random wormhole would also be unlikely to be traversable, meaning it's not big enough to travel through, or if it is big enough it's not likely to be long-lived enough to allow a return trip. Applying sufficient energy to the throat of the wormhole ought to fix it in place and make it traversable, calculations show, and zero point energy (ZPE, see Section 2.1) could supply that energy. (One calculation indicates that mass-energy on the scale of the mass of the planet Jupiter would be needed to stabilize the throat of a traversable wormhole. Such a level of energy could be readily available with ZPE.)

But even if rendered traversable, a random wormhole would still be in a random location, connecting to another random point, of no use to non-random travelers. Basically, if we are planning to use them for traveling we would need to construct them in pairs. In theory, you could do that by creating quantumly entangled black holes and then pulling them apart and placing the two termini at convenient locations. (Manmade black holes are not impossible, as explained in Section 5.3.)

But the necessity of installing them implies that we have ready access to both ends, which shoots down the rationale for using them to enable convenient interstellar travel. If we have to ship one end to a distant star, which could take centuries, what's the advantage of having a wormhole?

Actually, there could be several. If we have access to warp-drive starships as laid out in the previous two sections, we could use one to move an access point of a wormhole to the desired location. That might seem pointless if we're already operating faster-then-light starships and could make the trip without a wormhole, but remember, in a society where nearly everyone has access to an automobile, trains still operate, moving freight. Setting up wormhole links between high-volume access points might likewise be desirable, for an interstellar civilization conducting interstellar commerce.

But manmade wormholes might be even more important in the absence of faster-then-light starships. If you have one end of the wormhole at home and the other in a rocket heading toward a star, you could use the wormhole to keep feeding fuel to the rocket. The straitjacket of the rocket equation (as laid out in Section 4.0) goes away, simply because you don't have to accelerate your unused fuel.

Thus freed, you should be able to get your rocket well past half the speed of light, to the point where relativistic effects make further acceleration unproductive (as explained in Section 6.0.) At those speeds trips should take years and decades, rather than decades and centuries, and anyway the crew could be rotated in shifts to the outgoing rocket, making flight time simply an engineering issue. And, once it has arrived and been installed, further trips to that terminus would be immediate.

If another civilization were operating a wormhole, even in our vicinity, it's not sure how we would notice—except that visitors would keep showing up without any apparent difficulty, as is the apparent case. The openings might or might not emit distinct radiation. The black hole would not necessarily produce any gravitational distortions, since you're talking about masses that might be equivalent to multiple aircraft carriers, rather than the equivalent of solar masses.

In the end, wormholes sound like they could be too good to be true. But, of course, at one point we might have said that about the steam engine, electricity, and the personal computer. As to whether wormholes will join them, time will tell.

❖ ❖ ❖

6.4 – Faster-Than-Light Travel: Quantum Entanglement

In Section 6.0 we discussed the fact that (inconvenient as it is for interstellar travel) nothing with mass can ever reach c (the invariant speed of light) much less exceed it. Photons and other massless particles move at the speed of light, but also never exceed it. Basically, nothing can travel faster than c, with the implication that no information or influence of any kind can ever travel faster than light.

Except that there is direct, experimentally confirmed proof to the contrary, teasing us with the possibility of an exploitable loophole in the law of c.

It turns out that, in quantum mechanics, two or more particles can be entangled, meaning that their physical properties (spin, position, momentum, polarization, etc.) cannot be described without knowledge of the properties of the other members of the entanglement. It's as if their identities are blended. (There are numerous ways to create entanglement, and entanglement has been demonstrated in photons, electrons, neutrinos, large molecules of carbon, and even small diamonds at room temperature.)

As an example, let's say two particles are emitted by an atom during radioactive decay. They will be entangled, and their up-down spin (to name one property) will be complementary: if one has the up-spin property, the other will have the opposite down-spin property.

So far, so good—but this being the quantum realm, the particles will not actually have a spin direction until that direction is measured. (In fact, the values of all their properties will have only potential existences until they are likewise measured—but there are other theories about this, as seen in Section 7.3.) So when the spin property of one of the entangled particled is measured, it will be seen to have a spin direction (up, let's say)—and then the other entangled particle will take on the complementary spin direction (down, in this case.) It's as if the two particles consulted with each other concerning what roles they would take—except that the selection happens instantaneously. And it happens instantaneously no matter what distance separates the entangled particles. If they're too far apart for light to travel between them during the selection, it doesn't matter—it still happens instantaneously. In fact, experimenters can't tell which particle goes first.

Yet, the transmission of information in this universe is supposed to be limited to the speed of light. But that's not the case with quantum entanglement.

Dissident physicists long maintained that we must be missing something, that the particles must have "hidden variables" containing the information they'll need to make the right choice when any member of the entanglement is measured. There was a torrent of clever experiments trying to identify these variables and their operation, and perhaps finally expose the underlying mechanics of quantum mechanics and snag a Nobel Prize—with no luck. Apparently, what we see is what we get, there are no hidden variables, and the thing that Einstein called "spooky action at a distance" does happen.

But purists remain satisfied that the laws of the universe are upheld because, when measuring a property of a particle, you can't specify what value you want assigned to that property. If you could, you would force the other particles in that entanglement to adopt specific values for that property, and that would allow you to set up a communications system that's faster than the speed of light. With the status quo, you're triggering the display of random results when you make a measurement, and that is not the same as sending information.

But something that amounts to information for the entangled particles is traveling faster than light between them, and the only reason it doesn't count as information for us is that we haven't yet figured out how to set individual quantum properties.

But the big picture is that we haven't a clue as to what's going on with entanglement—and by inference the rest of the universe. Whatever it is that's passing between the entangled particles, it acts like space and distance don't exist, implying that space is an illusion. In other realms, space exists as the expression of an interlocking network of rigorously defined physical laws, so that the behavior of anything traveling through space can be calculated with precision.

With entanglement, it's as if there's nothing to calculate.

This alternate expression of space begs for deeper examination—although investigating something that may be illusionary is always a challenge. It may be that the dissident physicists had a point—the mechanics of entanglement may depend on particles having hidden variables, except the variables may be hiding in plain sight rather than being hidden. For instance, what if particles have a location property, so that members of the entanglement know where each other are? If so, perhaps they would be made to join each other—after all, they act like space and distance don't exist. Since we've seen large molecules with entanglement, perhaps we would use their locating mechanism to send fuel to a rocket heading for another star, freeing it from the downward cycle of the rocket equation (explained in Section 4.0) since it doesn't have to accelerate its own unused fuel.

Or perhaps we could find a way for cargo and passengers to piggyback, using some kind of container whose exterior would be entangled. Clouds of entangled particles left over from the Big Bang are very likely adrift in space, and we could jump from one selected particle to another. Different clouds, each with its own entanglement, doubtless overlap, so we could eventually get to any destination.

All this is speculation, and the fact that we don't know the bounds of what's possible illustrates the limited nature of our understanding of the quantum realm. We might as well be Michael Faraday (as mentioned in the Introduction) talking about hydroelectric dams when about all he really knew how to do was move puff balls with static electricity. About all we can say with confidence is a paraphrase of what he said: someday, we'll tax it.

❖ ❖ ❖

6.5 – Faster-Than-Light Travel: Heim Theory

It's called the "theory of everything" and theoretical physicists have been searching for it for nearly a century, hoping to combine quantum mechanics (with its explanation of subatomic particles) with relativity (with its explanation of gravity.) Neither can account for all observations. But if we had a successful ToE we could write a catalogue of physical constants and fundamental forces and see at a glance how they interact—and how to exploit those interactions for, among other things, spaceflight.

Except that an eccentric recluse in Germany may have done so in the 1950s, calculating, among other things, that you could travel faster than the speed of light with considerably less energy than is usually found in a single AAA battery.

Theoretical physicist Burkhard Heim (1925–2001) got his reclusive reputation the hard way, from a wartime accident that turned him into a nearly blind and deaf human lobster. At age 19 he was working in a German military explosives lab during World War II (he'd been drafted into Hitler's air force) when an accidental explosion cost him most of his sight and hearing, and both hands. The stumps of his forearms were then refashioned into so-called Krukenberg hands, using a procedure developed in 1917. The procedure involves the forearm's radius and ulna bones being separated and wrapped in skin, becoming prongs or long jointless fingers. Various muscles and tendons in the forearm are then reconfigured, so that the prongs come together when the muscles are clenched. Each forearm becomes a pincer, and the victim can manipulate objects while retaining a sense of touch and body position that is not possible with a prothesis, especially if the victim is visually impaired.

Recovery took more than 50 operations and Heim found that concentrating on theoretical physics was soothing, and that he could memorize formulae as they were read to him. His visual impairment, however, led him to develop his own system of notation for his calculations, cutting him off from the systems used in the mainstream. He was able to get an advanced degree and work at the Max Planck Institute, and consulted for the Glenn L. Martin

Co. (now part of the Lockheed Martin aviation conglomerate) and probably others.

But little of his work appeared in peer-reviewed journals. He continued refining his ToE until the end the end of his life, but stopped working on spaceflight around 1959.

As for Heim's ToE, it calls for three "real" dimensions (the familiar length, width, and depth) where things happen, plus time, plus two other imaginary dimensions if you want to calculate gravity and electromagnetism, plus two additional imaginary ones if you want to calculate the strong and weak nuclear forces. (While thus covering all four fundamental forces, his math predicts two further fundamental forces, which, if valid, may be associated with dark matter and dark energy. See Sections 2.3 and 3.4.)

While Heim allows for eight dimensions, the top four (i.e., the imaginary ones) contain what he called "steering coordinates" rather than objects or space. Their contents have been compared to the blueprints of a house, which are necessary for its existence, but not sufficient for its existence. This compares to string theory, whose extra dimensions are assumed to be real, but infinitesimal.

Under Heim's ToE, space, matter, and action have quanta (i.e., minimum quantities, so that they are not infinitely divisible), and space has significant physical features. In fact, space is made of a lattice of quanta that Heim called metrons, each with an area of 6.15E-70 square meters. Comparing a metron to a proton is nearly commensurate with comparing a proton to Planet Earth. But the fact that space is quantized rather than infinitely divisible means that gravitational attraction falls to zero at about 150 million lightyears— beyond that, the force falls below the minimum quantum.

Quantized space also implies that black holes as singularities should not exist, since infinitely small points can't exist. (Still, shrinking a solar mass to metron scale might accomplish the same result.) Heim also calculated that the universe is currently 5.37E109 lightyears in diameter. The universe began 5.45E107 years ago and originally consisted for three metrons, each about a meter in diameter. Matter began to form "only" 15 billion years ago—there was no single Big Bang—after the original metrons broke down into enough tiny ones to differentiate into matter. Occasional gamma ray bursts spotted by astronomers may be side effects of residual matter creation. Most red-shifting, meanwhile, is caused by gravitation rather than the expansion of the universe.

As for spaceflight, Heim calculated that electromagnetism can, under certain conditions, be converted into a form of gravitation, and emitting it from a spacecraft would actually reduce the inertial mass of that spacecraft. This would also increase the value of c (the speed of light) for the spaceship, dropping it into a spacetime where c has the new value. That done, accelerating past the original value of c should not involve relativistic effects, and if the mass of the spacecraft could be reduced 10 percent it should be able to go ten times original c. As soon as the spacecraft stops emitting the gravitation field it should return to the original spacetime.

Doing that to a spacecraft of 10,000 metric tons (comparable to a guided missile destroyer with a crew of about 300) that is in zero G would take 0.9 joules, or less than one watt-second of energy. (One AAA battery, off the shelf, officially contains 5,071 joules.) However, lifting it off the Earth's surface would take 625 billion joules. That's about 300 times the output of Hoover Dam, but would be readily available if zero point energy is realized (see Section 2.1.)

If Heim theory is not just a mathematical recreation and can be reduced to practice, and other civilizations have perfected the resulting technology, than the feats we see visitors performing in our skies become borderline trivial. Interstellar travel does not pose the danger of wrecking the Solar System (see Sections 6.1 and 6.2) and would be convenient enough to support commerce, with trips lasting months and years, rather than years and decades.

But Heim's ToE has not been accepted as the foundation of theoretical physics. It has no single feature that invalidates it, but there is no single feature that validates it, either—but there can be no such single feature. Yes, it can be used to correctly predict certain attributes of subatomic particles, but like a lot of theories it appears to be complicated enough that you can play with the numbers until you get a result you like. If they find Heim theory reliable, scientists will gradually learn to trust and use it and then develop technology based on its predictions. That will not happen overnight. But, of course, nothing lasting ever does.

❖   ❖   ❖

6.6 – Faster-Than-Light Travel:
Superfluid Vacuum Theory

A vacuum, of course, is the "thing" that's left over after everything else is removed. It's easy to say that's there's nothing there at all, and skip on to the next topic, except that there is space there, and it exists within time, so we still have spacetime. Under both relativity and quantum mechanics, empty space keeps reacting to events, so it's tempting to say it has properties. And if its properties are those of a superfluid, there's almost got to be a way to travel faster than light.

A superfluid behaves like isotopes of helium chilled nearly to absolute zero, becoming a fluid with no viscosity. It can only be stored in a fully sealed container, as it will happily flow up the inside wall of any other container and then down the outside wall, etc.

More importantly, waves flowing through a superfluid medium will be almost unimpeded yet will remain affected by the laws of fluid dynamics. That means photons with different energy levels will move at different speeds, and light waves will experience dissipation over time. The implication is that c (the speed of light in a vacuum) is not invariant, since the velocity of photons is no longer invariant. But the bedrock assumption of relativity (see Section 6.0) is that c is always the same, regardless of your velocity, your location, your direction of travel, etc., and that nothing with can ever go faster than c.

If the superfluid vacuum theory is true, the bigger implication is that spacetime, with its relativistic properties, is an emergent phenomenon of underlying components or processes that are themselves not relativistic—we just don't know what those underlying components or processes would be.

(Emergent phenomena involve systems whose properties are not evident in the properties of their constituent components. For instance, individual water molecules exhibit only molecular properties, but when brought together in large numbers, under the right conditions of temperature and pressure, they exhibit the emergent properties of water, with surface tension, hydraulics, etc. Under other conditions the same molecules will exhibit the (completely different) emergent properties of water vapor, steam, or ice. According to this analogy, we may be beguiling ourselves with the hydraulics of the lake of spacetime that surrounds us, when we should be trying to determine the properties of its water molecules.)

Meanwhile, if relativistic spacetime is an emergent phenomenon of underlying (and heretofore unknown) components and processes that are not themselves relativistic, that would mean that relativity is just the average behavior of spacetime under average conditions in our corner of the universe, and it could behave differently under different conditions. And so if we could define those conditions, and determine how to change them them, accelerating past c might not be out of the question. Relativity would still be a factor, and its effects would still be built into the GPS navigation system, for instance. But when it comes to star travel, accelerating past c would be a matter of controlling some spacetime factor that the physicists have yet to identify.

Concerning the number of ifs that would need to be fulfilled to achieve that scenario, no one's holding their breath, and it may seem odd that anyone is taking the possibility seriously. But theoretical physicists can dream too, and what they dream of is nailing down a Theory of Everything (ToE) with their names on it, and superfluid vacuum theory might be a plausible means to that end.

As noted in the previous section (and elsewhere) theoretical physicists have been searching for a workable ToE for nearly a century, hoping to combine quantum mechanics (with its explanation of subatomic particles and the behavior of electromagnetism, the nuclear weak force, and the nuclear strong force) with relativity (with its explanation of gravity.) Both have been thoroughly proven. But neither can account for all observations and they contradict each other in some settings. But if we had a successful ToE we could write a catalogue of physical constants and fundamental forces, giving us a cosmological version of the table of elements. This would not make us masters of the physical world—if it's like the table of elements, there will be no end of complicating conditions and special interactions. But it should put us in a better position to become a spacefaring civilization.

There are multiple approaches to a ToE using a spacetime emerging from superfluid. At least one would allow particles with mass to reach c (but possibly not exceed it) using finite energy. But all of them attempt to understand gravity through quantum mechanics, which operates at sub-atomic scales. In that realm experimentation is difficult and data hard to collect.

Doubtless someday we will crack the code, so to speak, and see if superfluid vacuum theory is a route to the stars, or is another branch of mathematical recreation. Until then it is another monument to our limited understanding of the cosmos—and another challenge to wrestle with those limitations.

❖ ❖ ❖

6.7 – Faster-Than-Light Travel: Hyperspace

As laid out in Section 6.0, in the spacetime we inhabit we can't exceed the speed of light, known as c. In fact, even reaching the speed of light appears impossible—unless we do some serious tinkering with spacetime. Otherwise, convenient interstellar travel is simply not possible—yet the evidence we see in the sky indicates that it can be done.

But it may be that spacetime, as it exists, has some backdoors that we could use to accomplish convenient interstellar travel. The clue may be in the common definition of spacetime, of three dimensions plus time. The first three you're familiar with: height, width and length. You can see them and move around in them and have no doubt they exist. The extra one, the fourth, is time. You can't see it or feel it, but you're probably open to the idea of accepting it as a real thing, as events keep happening in a sequential fashion that facilitates perception. Also, you keep noticing a time-related phenomenon called causality, where an event that appears to cause another event invariably precedes that other event.

So, if we're up to four dimensions, why not five, or more? For that matter, why can't the laws of physics in one of more of the extra dimensions be more amenable to interstellar travel?

Science fiction writers have long embraced this approach to exceeding c. Extra dimensions, often called hyperspace, are introduced to conveniently connect a character's point of origin and destination, with an explanation that relativity doesn't apply in those dimensions, or that the extra dimensions don't contain distance (that being left in the original three dimensions, after all), or that it just works somehow. In any event, the character gets to the destination in a way that the reader can accept, and the plot can continue unfolding.

Unfortunately, in the real world we have not identified anything like hyperspace—yet. But not for any lack of trying.

Basically, professional theoretical physicists, writing in peer-reviewed journals that are not read for entertainment, have also been busy inventing extra dimensions. Typically—but not always—

the extra dimensions are added to models of spacetime to explain why gravity is so much weaker than the other fundamental forces (electromagnetism, the strong nuclear force, and the weak nuclear force), a question otherwise known as the hierarchy problem. (It's often assumed that gravity is diluted by the need to propagate through all the extra dimensions, while the other three forces only propagate through the original three spatial dimensions, as they are too short-ranged or operate only by line-of-sight.)

For instance, various versions of string theory postulate 10 dimensions, 11 dimensions, or 26 dimensions. But you're not going to use them for travel, as these dimensions are assumed to be physical but are also assumed to "compactified" to the Planck scale, comparable to the Planck length of 1.6E-35 meters. That makes them about 20 orders of magnitude smaller than a proton's diameter (about 10E-15 meters.)

Various "universal extra dimension" models, which assume that gravity propagates through all dimensions, envision extra dimensions with a plumper size of 1.0E-17 meters, which is still too small for a proton to enter.

The size of the extra dimensions in the "large extra dimension" models depends on the number of dimensions to be added, but if the number is kept at two the dimensions can be millimeter-sized. That would at least admit protons, but you won't fly your starship through it.

Models with multiple dimensions of time have also been proposed, so that the flow of time does not have to be universal and absolute. How their availability would help space travel is unclear. And of course, the extra dimensions in Heim theory (see Section 6.5) are imaginary and therefore of no direct use in space travel.

The method used to enter these extra dimensions is also usually glossed over, but there is typically a not unreasonable assumption that significant energy would be required. That's also usually the case in science fiction treatments, with the energy requirement conveniently weeding out the riffraff, limiting access to properly equipped heroes and villains and other focus characters, while keeping the plot tightly focused.

The big picture is that the real-world theoreticians may be becalmed or making great strides in their sea of numbers, but at no point have they spotted anything that looks like the hyperspace of fiction. That may change, as the theoreticians achieve new insights and digest new experimental results. But in the meantime, hyperspace is simply too good to be true. It looks like it belongs where it came from, in the world of fiction, and we will have to look elsewhere to get to the skies.

❖ ❖ ❖

7

Time Travel: The Problem

This section is approximately 1,200 words long. If its content intrigues you, your subjective experience, upon completion, will be that it was quick and easy to read. Otherwise, you might subjectively remember it as lengthy and arduous. For an objective measure you might have timed yourself with a stopwatch. At the average rate of 300 words per minute for an adult with native fluency in English who is reading English-language material for comprehension (don't skim!) you should have finished in about four minutes.

But whether your approach was subjective or objective, you will only have measured time. You will not have defined it.

If pressed to define time you might say that time is the interval between events. But then you'd have to define the interval between events, and the definition of that is time. So we're stuck in a loop—yet we all know from personal experience what time is. Don't we?

In the International System of Units, the second is the foundational unit of time. The second used to be defined at one-86,400th of a day, there being that many seconds in 24 hours. But the length of a day can vary, so now, officially, a second is defined as 9,192,631,770 vibrations of a caesium-133 atom. The measurement is made with clocks that are said to be accurate to one second in 30 million years. Physics also recognizes an interval called Planck time as the shortest unit of time that could in theory ever be observable. It's the time it would take light to travel a Planck length, equal to the wavelength of a specific quantum particle, or about 1.6E−35 meter, and amounts to 5.391E−44 second. However, most theories do not quantize time, seeing it is infinitely divisible.

But look again—these too are measurements, not definitions.

Leading scientists and philosophers have tried to get beyond circular definitions, but have disagreed about basic issues, like whether time is, for instance, absolute and universal (as Newton thought) or is relative to motion (as Leibniz thought) or an artifact of our perception and instincts (as Immanuel Kant believed.)

Today, we still lack a good definition of time, and despite it being integral to science, time remains something that we can't isolate in a lab, attach probes to, and subject to tests and experiments. Meanwhile, calculations in physics and non-organic chemistry are indifferent to time and will work fine forward or backwards. In fact, no physical law appears to require that time run in any particular direction, or at any basic rate, or at all. Nor is there any obvious reason why all sentient beings appear to experience "now" simultaneously. (Your perception of the on-going present is, incidentally, constructed by your brain rather than directly sensed, and probably covers a span of about two-tenths of a second.)

It gets worse. It turns out that the two well-established and thoroughly proven theories that dominate theoretical physics, quantum mechanics and relativity, are built around time, yet fundamentally disagree about the nature of time, to the point of direct contradiction. This remains the case even after generations have been spent trying to reconcile the two in order to arrive at a Theory of Everything.

In quantum mechanics, time is an absolute, immutable, global parameter, one that is external to whatever system is in question. The source of time is not a consideration, its flow never changes, and there is no evident way to change it. Consequently, when (for instance) a quantumly entangled particle is measured so that the status of its properties transform from potential to real, the other particles in that entanglement will undergo complementary quantum transformations at the exact same moment, regardless of their relative velocities or distances from each other. If the magnitude of their separation was large enough that the speed of light should have made such simultaneity impossible, too bad—the transformations were simultaneous anyway.

With relativity, time is formulated as the fourth dimension and while inescapable is not fixed and absolute, with the flow of time depending on the observer's frame of reference. Clocks on the ground move at one speed, those in motion move at a slightly slower speed, and those moving at nearly the speed of light move at significantly slower speeds. The lack of absolute time makes it impossible to determine if distant events were simultaneous (and therefore contradicts quantum mechanics, whose functioning demonstrates absolute time and simultaneity.) However, events that were separated by time in their original frame of reference will never appear to be simultaneous in any other frame of reference, and cause-and-effect sequences will appear to remain intact. Meanwhile, the speed of light will be the same in all frames of reference, and in all known cases light travels forward in time. (The theoretical tachyon particle that moves faster than light would supposedly travel backwards in time. But no tachyons have been discovered yet, nor is there any agreement that they are even possible, or would be discoverable if they existed.)

Clearly, our understanding of time is comically inadequate. We can't agree on a basic definition and our two leading cosmological formulations fundamentally disagree about the nature of time, even while otherwise being successful at predicting real-world events.

Section 6.6 discussed the idea of emergent phenomena, whose properties are not discoverable in the properties of the fundamental phenomena they emerged from. As said, the properties of hydraulics are not evident in the properties of a water molecule, but if you get enough water molecules together under the right conditions, you have hydraulics. Time may also be an emergent phenomenon, and our attempts to understand it are like an attempt to understand hydraulics in the absence of any knowledge of water molecules.

Meanwhile, the fact that we don't see any time travelers among us, presumably wearing loud shirts and clicking their cameras at a still intact landmark while loutishly complaining that their insurance carrier would not allow them to visit on the day of the landmark's dramatic demise, has been used as an argument that time travel is not possible.

But we see visitors overhead, and we can be reasonably confident that they came here from a long way off, since our near-field surveillance abilities keep improving. But if they came a long way, there is no guarantee they came a long way in terms of physical distance—they could have come a long way through time (or, admittedly, through both space and time.) Assuming they're from millennia or even eons in the past or future, cause-and-effect paradoxes would not be an issue, since changes they make here would not be evident at their place of origin, thanks to its lengthy detachment in time. And while here they would have every reason not to exit their conveyances and take selfies—if the past is a foreign country, so is the future, and both would harbor germs they would want to avoid. But our world might still offer other, specific, predictable, resources they might want to reap, making the trip worthwhile. (For all we know, the DNA of our algae could be priceless to them, etc.)

That said, when we talk about space travel, it goes without saying that space, celestial bodies, their properties, and the distances between them all have an objective existence that is independent of our existence, or of our velocity, or of any emotions we feel toward them. We know we can travel in space to other celestial bodies, even to distant stars, and the only question is how to make that travel convenient to us. The same cannot be said about time. Its properties remain a mystery to us and its very existence is questioned. That being the case, the question of traveling in time involves a lot of assumptions—and the commonest and simplest assumption is that time travel is not possible.

The following sections explore sets of assumptions which would make time travel possible. Yes, the big picture is that there are serious scientists who accept the possibility of time travel and foresee ways it could happen. But while we're in big picture mode you'll notice that these sections contain a lot of assumptions and very few discrete facts. Before we journey to other eras, that will have to change.

❖ ❖ ❖

7.1 – Time Travel: Looping in the "Persistent Illusion"

Einstein once famously characterized the difference between past, present, and future as a "stubbornly persistent illusion." But this does not mean he dismissed time as unreal and illusionary—the quote is from a letter of condolences to the family of a departed friend, affirming Einstein's belief that he would soon be joining the departed.

No, Einstein's relativity equations treated time as real and measurable (albeit again without a definition), subject to real distortions when stressed and especially dilation when accelerated. But Einstein himself was reportedly startled to learn that the implications of his equations allowed for the creation of a form of time machine, and a generation of physicists have struggled to come up with ways to head off the paradoxes that might arise from using it.

The proposed device is sometimes called a Thornian time machine, after American theoretical physicist Kip Thorne, and it differs from a typical Hollywood late-show time machine in two major ways.

The first is that a Hollywood machine lets the hero dial up the desired destination date, be it in the past or future, then step out into that setting and start having plot-advancing adventures—and then, if things don't go well, do some frantic dialing and return to the point of origin. But a Thornian time machine only lets you travel to the future and then loop back to the approximate point where you started—without stopping anywhere along the way. The loop will take a perceptible amount of time to traverse, proportional to its length, and the time traveler will experience the passage of that time. But upon stepping out of the machine on arrival, no time will have transpired at the point of origin, and the crowd gathered at the door will express surprise that the chrono-naut has aged in the blink of an eye.

Since the time traveler will only have interacted with the interior of the time machine during the trip, a Thornian time machine might seem a lot less useful than the Hollywood version. But consider: what if you had a computational problem that would take a computer years to calculate. In this case you could put the computer in a Thornian time machine, dial in a big enough loop, press the GO button, and when the door reopened a moment later the computer, now a trifle dusty, would be sitting there with the answer on its screen. (This feature is called "negative lag." Be sure and ask about it next time you're talking to a computer salesperson.)

The second major difference between Hollywood and Thorne is that the Thornian machine is based on science—or, at least, a widely accepted views of the relationship of relativity and spacetime.

This view is usually expressed using diagrams involving so-called light cones extending up and down from a horizontal plane representing the present. The point where they come together on the plane-of-the-present represents an event or object in a 3-dimensional space. Time is assumed to be moving from the bottom of the diagram to the top. Points below the horizontal plane are in the event's past, while points above it are in the event's future.

Points below and above the event but inside the cone are said to be in the event's light cone, since light from those points might (if in the past) reach the event and be involved in its causation or (if in the future) be reached from the event with the possibility of causation linked to the event. Points outside the cones have no connection with the event. The sub-lightspeed movement of an event or object from the past through the present and into the future is called its worldline. Worldlines are not straight since they follow the motion of the subject in both space and time. The worldline of an object that stays inside the cones is called a time-like curve.

Meanwhile, with carefully chosen parameters, it is possible for worldlines to form closed time-like curves (CTCs) which loop back on themselves, allowing the previously mentioned Thornian time machines.

The fact that a Thornian time traveler could arrive at the end of his loop-trip and kill his younger self before he gets aboard the machine has alarmed theorists, who have scrambled to formulate self-consistency or chronology protection principles for spacetime in order to save the universe from the resulting paradoxes. What does it mean to spacetime if the cause of an event was in the event's future, rather than its past? For that matter, is causation real, or an illusion? Would such paradoxes really pose a threat of some sort to spacetime, or just be shrugged off as noise by bystanders? No firm answers have surfaced, but on the other hand no one has proposed a practical way to build a Thornian time machine.

However, the discovery of CTCs by respected theorists has forced the scientific establishment to take time travel seriously. But accepting the idea that the visitors we see in the skies are using Thornian time machines is a little harder, since such machines only allow you to move from A to A while aging, while you'd expect time travelers to prefer to move from A to B and then back to A while not aging. Also, we don't know what physical plant a Thornian time machine would need to generate the necessary CTC in spacetime, so we don't know what side effects to look for—fireworks, EMPs, visual distortions, etc.

Suggestions for generating the necessary spacetime distortions, based on our understanding of spacetime, have included a dense spinning cylinder that's infinitely long (wrapped into other dimension, presumably), a spinning black hole, a customized wormhole, or the mass of Jupiter crammed into a sphere five meters wide. None of these things could be operated in Earth's vicinity without humanity noticing something.

But, as mentioned in the previous section, our knowledge of time is laughably limited. Every single, distinct fact we can confirm about time will significantly increase the sum total of our knowledge about it. Perhaps, after we get a few such facts confirmed, it may become clear that it is possible to make a Thornian machine that goes from A to B before returning to A, and that visitors are clearly doing so. Or it may become clear that such machines are impossible, visitors are not using them, and the whole concept is just another mathematical recreation.

Time will (actually) tell.

❖　❖　❖

7.2 – Time Travel: Around "the Block"

The past no longer exists, the future isn't real, so all we have is the present, which is, however, a pretty big place. The previous sentence is a simple, intuitive, satisfying world view, often called presentism. It's one that does not permit time travel, since the past and future are not there to travel to.

Except presentism is a world view that is contradicted by Einstein's relativity, which has time flowing at different rates in different frames of reference, shooting down the concept of simultaneous events (especially distant ones.)

The major alternative to presentism is eternalism, in which the past, present, and future all exist equally, stacked into what some call the block universe. The present is our collective attention somehow sweeping through the block in synch, so that time truly is an illusion and has no objective existence.

If the past and future both exist as much as the present, we ought to be able to visit them. The trick would probably be to leave the present, but perhaps a Thornian time machine (discussed in the previous chapter) modified to allow wider travel might suffice—our knowledge of time is so limited that we can't say if such a thing is possible. But we can probably stop worrying about the impact of paradoxes—if the block already exists, that implies that events are predestined, whatever happens will just happen, and there will be no paradox. If a time traveler kills himself so that he cannot kill himself, then it happens, and the universe will not have a nervous breakdown.

But if the future already exists and events are predestined, that means our acts in the present can't change the future any more than they can change the past. Likewise, there can be no such thing as cause and effect, as the block has no obvious connection between causes and effects in different layers (although some theorists would allow a "spacious present" that would accommodate some cause-and-effect relationships.) But that also makes the block compatible with both relativity and with quantum mechanics (and with Aristotle, for that matter) since what happens is what happens, regardless of whether the two theories are contradictory, and no further explanation is required. If two distant but entangled quantum particles experience a wave function so that they collapse at the same instant (as required by quantum mechanics in a manner contradictory to relativity) it was predestined to happened. If time dilates in a particular way in an accelerating frame of reference (as required by relativity in a manner that ignores quantum mechanics) it was predestined to happen.

As for the bigger question of why anything at all happens, here we must again consider, as elsewhere, the question of whether our consciousness is a product of the universe, or the universe is a product of our consciousness. If the universe is a product of the collective consciousness of its perceivers (i.e., you, your cat, members of alien civilizations, etc.) then the events in the block were presumably agreed upon by some kind of Committee of the Whole, and there are presumably other blocks where other courses of events are being explored. Our superficial personas that live in the here-and-now experience life in the block, but the behavior of those personas does not determine the already existing future layers in the block, so they're off the hook in terms of getting what they deserve.

But don't get too comfortable—there's an alternate version of eternalism called the growing block universe, in which the future is not predetermined, but grows from the present and is then added to the past. In this cosmology, we're stuck with the results of our actions, and can't entirely blame fate. (But apparently some things remain predestined—how else can we explain the radical difference in the fates of the residents of 1945 Hiroshima versus those of, for instance, 1945 Detroit?)

The growing block universe implies that the past remains real and could be visited, whereas the future doesn't exist and therefore can never be visited. That would seem to fly in the face of the Thornian time machine, discussed in the previous chapter, but actually does not—that machine lets you loop into the future but not visit it before returning to the present. On the other hand, no equivalent conceptual framework for visiting the past has yet arisen.

Paradoxes arising from visits to the fixed past, if such visits prove to be possible, will presumably not be an issue. If the past is by definition fixed, a visit by us will not change it, and the implication is that residents of the past will not be able to see or interact with visitors from their future, or be exposed to any information they might carry, including the fact that they exist. Visitors to the past, meanwhile, will only be able to bring back information—i.e., report what they saw.

In the end, our knowledge of time is so skimpy that we can't say for certain if eternalism or growing block is the operative cosmology in our universe. On the other hand, we can't rule out either possibility (or most others.) But it does seem safe to say that neither cosmology would seem to promise worthwhile forms of time travel, such as could form the basis for interstellar commerce. Additionally, if visitors were using either approach, we would have no way of knowing it, as we would not know what collateral effects (shock waves, flashes of light or x-rays, etc.) to look for.

Presumably, that will change (assuming events are not predestined to turn out otherwise.)

❖ ❖ ❖

7.3 - Time Travel: Many Worlds

Schrödinger's cat is one of the most recognizable figures in theoretical physics, ranking perhaps only after Einstein himself, and the animal's fate has been furiously debated and recalculated since 1935. One of the results of that effort implies that time travel need not present dangerous paradoxes to the spacetime continuum. In fact, time travel might be little different from any other form of travel—but heaven help anyone trying to draw a roadmap.

It is not reported if Austrian (Irish after 1948) physicist Erwin Schrödinger (1887-1961, Nobel Prize in physics 1933) had a feline pet at the time, but he used a cat in a thought experiment to illustrate what he saw as a problem with a leading interpretation of quantum mechanics. Basically, the thought experiment would, if followed faithfully, produce the paradox of a cat that is simultaneously alive and dead.

He was worried about quantum superpositions, a fundamental principle of quantum mechanics, which says that a quantum particle (or entangled particles) can exist as a combination of multiple states corresponding to different possible outcomes. According to the then-popular Copenhagen Interpretation, a quantum system remains in superposition (otherwise called a wave function of probabilities) until it interacts with the external world, even just by being measured. At that point the superposition collapses into one of its possible definite states.

Schrödinger seemed to think the Copenhagen Interpretation was well and good with photons and atoms, but he worried that it was possible to set up macro-scale quantum superpositions, leading to ridiculous situations.

To demonstrate the problem, Schrödinger proposed a thought experiment involving a sealed box containing a cat (which we'll assume is napping), a flask of poison gas, a single radioactive atom, a Geiger counter to detect radioactive decay, and an actuator to break the flask. If and when the Geiger counter detects the decay of the atom the actuator is triggered shattering the flask, the poison gas is released, and the cat dies. Otherwise it continues napping.

So far, so good—but under the Copenhagen Interpretation of quantum mechanics, at some point the cat shares the superposition of the decaying atom, and is simultaneously alive and dead—and remains that way until a conscious observer opens the box to perceive its state, thereby performing a quantum measurement and collapsing the cat-atom wave function. Only at that point is the cat alive or dead.

But in the real world, where common sense rules, the cat will remain alive until it dies—there is no superposition of states.

Theorists have raced to restore common sense to the cat's fate, and new interpretations of quantum mechanics are often judged by their effect on the hypothetical cat. Hereafter we will examine the increasingly popular Many-Worlds Interpretation (MWI) which not only restores common sense to the cat's situation, but appears to have the potential for legitimizing time travel.

Under the Copenhagen Interpretation, there can be only one outcome of a wave function collapse, and after the collapse the cat must be either alive or dead. Under MWI, there is no collapse of the superposition of states, as both outcomes take place—in different, parallel universes that spring into being to embody the outcomes. In one, the cat will remain alive, and in the other the cat will be dead. Satisfying common sense, it will not be simultaneously both. Any quantum interaction that has multiple possible outcomes will have a similar result—universes will pop into existence to embody each possible outcome. These universes will not be in communication with each other, and each inhabitant of each universe will assume they are the rightful embodiment of themselves, remaining unaware of their alternate selves in other universes, who go on to their parallel but separate fates.

First formulated by American physicists Hugh Everett III (1930-1982), MWI implies that we reside in a multiverse throbbing with countless universes constantly budding off from each other as their quantum particles interact. Because of the quantum interactions of particles in the farthest reaches of the universe, not only are there are countless versions of you reading this article, but there are countless more from moment to moment. The interactions are not only measurements by other systems (not just sentient observers) but the decoherence of previously entangled particles.

Notice that there is no mention of the speed of light. An entire new universe, scores of billions of light years across, pops into being from one trigger point, the necessary information apparently conveyed instantly throughout the entire new branch universe from the entire old trunk universe. This is another example of the incompatibility of relativity (which restricts all forms of influence to the speed of light) and quantum mechanics (which allows instantaneous, literally universal influence.)

A common misconception of the MWI is that a new universe is engendered each time you make a decision, so that there is a universe where you turned this page before sipping your drink, and one where you turned this page afterward taking a sip. Branching continues with every decision, leading to situations that are more and more divergent. Since this has been going on since birth, every experience you'd reasonably have had has been explored, so there must be a universe where you're married to that person you had a crush on when you were 14, but who painfully snubbed you.

Actually, that's not what MWI says. Instead, MWI implies that the branching continues relentlessly because quantum mechanics demands that it continue, regardless of any decisions the inhabitants may be making. Most of these branch universes will be identical to their previous trunk universes save for one quantum outcome. As the branching continues, divergences will appear as the inhabitants make slightly different decisions based on slightly different circumstances. So you may take a sip before or after finishing the page, depending on many shifting factors. As time goes on, the divergences will become significant, and the inhabitants will have widely different experiences—but that does not guarantee that every possibility will be explored. The adolescent decision-making process in force when you were 14 would probably still have led to your humiliation. But since branching has been going on since the beginning of time, there could be divergences among your ancestors so that you and the other person never meet, or you meet but that person will have different genetics, or you do, etc.

But that's not why MWI is attractive to would-be time travelers. There are actually two reasons, both based on practical considerations.

The first is that, in theory, the universes that MWI engenders are not supposed to be in communication with each other. Once created, they go their own way, although meanwhile themselves constantly creating new branches. But for the branching to happen, there must be some kind of communication between the trunk and branch universe, or the branch universe would not know what it contains or what quantum alternative it embodies. This leaves open the possibility of navigating from one to another, perhaps even working backwards and forwards. Mapping the thicket of endless, nearly identical universes would be challenging, but that's the kind of thing computers are good for.

The second is that the so-called grandfather paradox, which greatly bothers theorists, would just go away. The theorists worry that if you went back in time and killed your grandfather, the resulting paradox could be catastrophic. After all, where did you come from? But with MWI you carry your personal history with you, wherever you go in time. If you move to a timeline containing your grandfather and kill him, that is just one event in one of the foaming masses of timelines that contain your grandfather. The fact that you were able to do it indicates that you were born in a timeline where your grandfather was not killed. Branching after the moment of the assassination will be of timelines where your grandfather is dead and you are alive. But there will be myriads other parallel timelines where your grandfather is untouched and goes on to have you as a descendant, and you go on and travel to the fatal timeline.

Keep in mind that MWI has not been experimentally confirmed, and may not be confirmable with current or foreseeable technology. That we're happy to envision a multiverse foaming with clones of ourselves because we're unhappy with the hazily understood fate of Schrödinger's cat is scandalous to some theorists. Nor does MWI seem to satisfy Occam's Razor, which demands that explanations be as simple as possible. But of course, sometimes complicated explanations are the correct ones—the entire field of organic chemistry is an example.

If MWI holds true and works as described, the visitors we see in the sky could as easily be coming from other epochs as from other stars—perhaps more easily. And, as if often the case, they would seem only slightly interested in us. After all, if they're from the future they already know all they need to know about us. If they are from the distant past, they know they will be a lot of branching between their time and ours, potentially leaving few commonalities.

◆ ◆ ◆

8

Non-Technology Alternatives:
Overview

The underlying assumption throughout most of this book is that we are products of the universe, which exists independent of us. By understanding its laws and methods we can become masters of increasingly larger domains, but if we kick a rock that foot will suffer, because the rock is real.

What if that's not the case? What if the universe is an expression of our inner selves, but the fact is largely missed by our surface personas, bound as they are to the illusions of time and causality?

If so, frankly, not much changes. The pursuit of science will still lead to greater conveniences, broader control of the physical world, and less unnecessary suffering. Kicking a rock will still hurt. But there will always be a veil that science will not be able to part, because science is about the physical universe, and the physical universe is not everything.

But if you accept that the universe is an expression of the inhabitants, you might feel it's more inviting to turn your back on physical things and the arbitrary agitation they demand, become a hermit, and pursue inner peace. There have always been people who have made that decision, even before tasteless advertising and mass consumerism prevailed. But most people nevertheless do not become hermits, and instead concentrate on the agitation needed to acquire physical conveniences. They seem fond of reliable food, safe drinking water, honest police protection, reliable medicines, etc. And indeed, having those conveniences surely removes many obstacles to the pursuit of inner peace. For a civilization to last, there presumably must be a balance between the material and the transcendental, either by individual choice or by institutional traditions.

It seems safe to assume that would be the case for any extraterrestrial civilization. They would be aware of the dichotomy between the physical and the non-physical, and their pursuit of science would have provided them with ample hints that science cannot explain everything. They, too, would have sought a balance between the two domains, although its expression may differ widely (as it does with humanity.)

A bigger question may be whether the fact that a civilization is "advanced" means it has mastered the physical world, or that its population has been able to find enlightenment and inner peace. In the latter case, we would probably not see them in our skies, since they'd likely no longer feel any compelling reason to go anywhere. (That their enlightenment would require them to travel across interstellar space to offer to help us is also unlikely, as they'd likely view unsought offers of help as intrusive and disruptive products of elitist self-congratulation. Our own historical experience with colonialism gives us plenty of examples of the disruption that results when a civilization with scientific achievements collides with one that lacks them, and any other civilization would likely have had an equally uneven history.)

More likely, any visitors we see in our skies would be from civilizations that are still investing in the pursuit of science—which makes the visits possible—while also recognizing non-physical factors. In fact, they may have invested in the pursuit of non-physical factors at least as much as they've invested in physical science, and consequently display what we'd called paranormal powers.

Admittedly, it also possible that they have turned their backs on non-physical factors, or are incapable of considering them. In that case, unless they were born with paranormal powers (perhaps endowed by evolutionary stress) they will not acquire any. The following sections are not about them. (Some visitors may, in fact, be inanimate, as noted in Sections 8.x.)

But more than likely, any civilization that has the surplus energy to devote to the exploration of science could also devote some to the exploration of non-physical factors—and that's been the case with humanity. The problem is that non-physical factors are apparently quite subjective and cannot be pursued with the tools used successfully in physical science, especially replicable experiments and mathematical analyses. That being the case, researchers who use the tools of physical science to probe non-physical factors would have little success and would lose patience—as has been the case with humanity.

But it doesn't have to be that way. A civilization that found ways to successfully probe non-physical factors (in ways probably idiosyncratic to them) could be expected to pursue them just like physical science, with comparable results.

The following sections are about such cases.

❖ ❖ ❖

8.1 – Non-Technology Alternatives: Telekinesis/Levitation

Whether the universe was produced by us as an expression of our inner selves (as considered in the previous section) or whether we are ourselves products of the universe and its physical laws is a question that is central to any consideration of telekinesis (also called psychokinesis), the power to move things with your mind. (Levitation is the power to do the same to yourself.) If the universe is our creation, telekinesis could potentially be innate—if we made the universe, surely we can move parts of it using our transcendental knowledge of its nature. But if we see ourselves as a product of the universe—creations of the same physical forces that we seek to control—then possibilities for the expression of telekinesis would appear to be far more limited.

Full conscious knowledge of the nature of the universe is probably not consistent with continued physical habitation of the universe, as its inescapable laws of physics place severe restrictions on personal behavior. Basically, anyone who could escape from those laws would surely do so. Consequently, if the universe is a product of our inner selves, transcendental knowledge of the universe would likely remain largely unconscious, even for a species that derives power and abilities from that knowledge.

If there are beings with largely unconsciousness transcendental knowledge, their behavior might resemble that of comic book superheroes or Ancient Greek gods, as their interaction with physical laws would be largely optional. At the same time, they would probably possess no conscious awareness of the mechanics of those interactions. They could fly or smash boulders because they decide to do so, while remaining indifferent to how they do it, much as you may decide to move a finger without detailed knowledge of neurons and muscles. As with yourself, the ability to move that finger, plus the rest of your body, requires metabolic functions, leaving you, and them, with various appetites and desires. But when it comes to flying or boulder-breaking, they would unconsciously invoke their transcendental knowledge to trigger a limited change to the universe around them, whose end result would be flight or boulder-breaking. Since there's no exertion of physical force, no physical laws are broken.

Such beings would be dangerous to be around, even if they resembled us, as they would share few of our limitations. In fact, it's hard to imagine what their agendas and desires would be—but as long as they have metabolisms they would surely have agendas and desires. On the other hand, it could be that their mastery of the physical universe leaves them with a drive to find inner peace rather than pursue trivial matters like material attainment. As a result, they would see no reason to travel. But if we see them in the sky that could mean they have inner turmoil similar to our own, and will have to be interacted with on that basis.

They could probably move at will between the stars without spacecraft, without regard for distance or relativistic effects, since they are altering the universe rather than traveling (as discussed in Sections 6.1 and 6.2.) However, they'd probably bring spacecraft along anyway, for personal comfort. They would move those craft using their usual powers.

Flashy, large-scale displays of telekinesis could be taken as evidence of a transcendental race of visitors, since even mass maneuvers would not require them to expend significant energy resources. Again, careful surveillance of visitors will be required for we could make any such determination.

Low-key displays of telekinesis for ordinary tasks (such as pushing levers) would be evidence of the other approach, where the users see themselves as products of the universe and totally subject to its laws—but those laws allow some form of telekinesis as a personal ability That's ironic, as various human psychics who have sought the spotlight claiming to possess the power of telekinesis promote themselves as products of the universe, yet possessing some ability to break its laws. Their feats, such as bending spoons and restarting failed mechanical wristwatches, invariably mirror conventional stage magic, with no applicability outside the stage.

Hundreds of actual scientific experiments under controlled

conditions have attempted to demonstrate telekinesis, presumably by people who likewise see it as a physical power rather than the innate product of transcendental consciousness. The experiments often involve trying to influence the results of thrown dice, or similar random events. Overall, the combined results of these tests slightly favor the existence of telekinesis. Meanwhile, this slight proof is itself cited by commentators as an example of publication bias, i.e., a result of the fact that experimenters are more likely to publish exciting positive results as opposed to mundane negative ones.

If a civilization (perhaps including ours) were to methodically establish and explore telekinesis as a physical ability allowed by physical law (as opposed to an expression of transcendental even if largely unconsciousness knowledge) the results might resemble magic, yet not violate the laws of physics. The user would harness some bodily control over one of the physical forces that can be exerted at a distance, and learn to direct it and use it to move or otherwise affect things external to the user's body. Certainly, it will be worthwhile to summon objects to you, etc. But, as those objects get larger and larger, you'll have increasing difficulty moving them, as it will take more and more energy. Moving a stuck car with your new power is not going to happen, as the car will weigh much more than you. If anything, you'll pull yourself toward the car. Strategically nudging aside an otherwise inaccessible obstruction might be an option, however.

As for bending spoons or restarting mechanical watches, you could use your new power do that, too—but you always could. Spoon-bending is a sleight-of-hand trick. If there's no opportunity to divert the audience long enough to impose a bend, simply holding a straight spoon in your slightly cupped palm may convince the inobservant long enough for the trick to work. Older mechanical watches, meanwhile, may get gummed up with congealed grease when left overnight at the bedside. Holding one in your palm will often warm the grease enough to allow the watch to restart.

But be sure and do it theatrically, as if you're exerting some mysterious force.

❖   ❖   ❖

8.– Non-Technology Alternatives: Dis-Incarnation

Of course, the most efficient way to travel is to leave your body at home and just send your mind to the destination. (There's a close second place, which we'll discuss later.) Being massless, your mind should get to its destination without any problems with fuel or acceleration or relativistic effects. You won't be able to bring back anything through customs, and so the practice can't serve as the foundation of a space-faring civilization, but for efficient scouting it has no equal—in theory.

In practice, the idea is more than a little controversial, even while its acceptance is surprisingly widespread. Mental travel goes under many names, such as astral projection, out-of-body experience, or remote viewing (or daydreaming, according to skeptics) but there is no shortage of anecdotal reports of it happening. What there is a shortage of are researchers willing to say that it has been demonstrated in replicable experiments or that it has been reliably harnessed to generate useful information.

That's not for any lack of trying, however. For instance, the US Army operated a remote viewing program from 1978 to 1995, apparently in an effort to catch up with the Soviets, who were reportedly getting results from psychics who could "see" what the Americans were doing. Despite this advantage, the Soviet Union collapsed in 1991 and the reports of its paranormal successes began to look like the usual disinformation. The CIA took over the US Army project in 1995 and decided that its few successes were indistinguishable from informed guesses, and shut it down. The British government also briefly had a program to study remote viewing, which shut down after no viewing could be demonstrated, much less studied.

Out-of-body experiences (OBEs) are less controversial then remote viewing, and show up in numerous undisputed anecdotes. Instead of spying on a distant country, the subject's viewpoint remains, typically, in the same room, but has moved outside the subject's body, which the subject is able to view. OBEs are typically triggered by severe physical or emotional shocks, or near-death experiences, making them difficult to replicate and study. The conventional supposition is that OBEs are a survival mechanism that lulls us into possum-like inaction during extreme situations. The victim's altered viewpoint is a distracting hallucination, rather than consciousness functioning separately from the body. Consequently, occasional reports of victims recalling an ability to move their viewpoints around at will, and waking with information they could only have acquired through remote viewing, are controversial.

Astral projection assumes that the soul of the subject leaves the body and travels to another place, basically accomplishing remote viewing. Many religious and cultural traditions include some form of astral projection, sometimes with elaborate descriptions of the process. Scientific evidence that astral projections are real remains lacking, as does scientific acceptance of the phenomenon.

With all these versions of remote viewing in circulation, it is tempting to assume there is something to it. As mentioned in the previous sections, an organized investigation of non-physical topics might produce results comparable to the results generated by the organized pursuit of science, but different, more idiosyncratic methods will probably be required. The failure of the US Army program might be seen as a wet blanket for that idea, but, reportedly, the participating remote viewers were not provided any feedback on the success or failure of their efforts, giving them no basis for improvement. In other words, no attempt was made to develop their talent—they had talent or they didn't, and if they had it the authorities didn't want to discourage them with bad report cards. Any organized investigation would need to give the participants the opportunity to develop their talents through some kind of feedback loop, so they could build on whatever success they might achieve. The process might resemble sports or games rather than science. The results would likely be different for each person. As ever, the replicable experiments and mathematical analyses that guide researchers in scientific investigations may have little applicability in non-physical investigations.

As for its potential, there is again the question, as mentioned in the previous sections, of whether we are products of the universe, or the universe is a product of our (largely unconscious) expression. If the latter is the case, then consciousness presumably can function separately from the body, and something like remote viewing ought to be possible via direct perception. But if we are products of the universe, that means our consciousness is the sum-total of the neuronal activity inside our skulls, and remote viewing will require indirect methods, such as telepathic perception from someone at the scene. (This would not involve the pitfalls of telepathy noted in the next section, as only visual images would be involved.)

If visitors are using such methods, there's no guarantee we'd know about it. If they are using direct perception, all they'd need is directions for locating their viewpoint, using coordinates they can relate to. Space being so vast, they'd probably be challenged to zero-in on coordinates that consistently lead to results. Using indirect methods, they'd need remote viewers with enough sensitivity to find well-placed viewing candidates.

But either way, remote viewing would not be as good as being there, since there'd be no frame of reference—you couldn't know the size or distance of what you're looking at without careful analysis, such as what the military uses for photo interpretation. That being accepted, it would certainly be worth doing. Consider the robotic probes that NASA sent to land on the moon prior to the Apollo landings. Fundamentally, there was no mystery about the proposed landing sites, as they could be seen from the Earth, but there was nothing like getting the view from the ground.

Meanwhile, we mentioned that there's a close second to mental travel in terms of efficiency. That would be teleportation, where your body is transmitted to its destination. Presumably, either you are converted to energy and transmitted for reconstruction on arrival, or you are disassembled atom by atom and the information transmitted to your destination where it will guide your reassembly using local materials.

If the idea terrifies you, it may be because you understand the complexity of the task, and the many things that could do wrong even if such technology were attainable. The Star Trek science fiction franchise has used teleportation as a plot device since its inception in 1966, but the writers have been vague about whether its "transporter" was supposed to be based on energy or information rendering. Either way, how (or why) the consciousness of the traveler would take root in the destination version of his or her body is unclear. But using teleporters require less screen time than landing a spaceship.

Of course, as mentioned in Section 6.4, transmitting liquid hydrogen to a traveling rocket, for use as fuel, could free it from the tyranny of Newton's Third Law, covered in Section 4.0, since it would not have to accelerate unused fuel. Meanwhile we're talking about transmitting a parade of identical, inanimate atoms rather than hoping to reassemble a living, conscious entity. Compared to teleportation, the technology is actually conceivable.

❖ ❖ ❖

8.3 - Non-Technology Alternatives: Telepathy

Telepathy implies a person projecting thoughts or feelings to the mind of another, or one person detecting the thoughts of feelings of another, without any use of the usual physical senses, such as sound, touch, taste, sight, and smell. Not included are stage magicians "reading" the thoughts of someone standing in front of them, or a poker player sizing up another player at the table. Those and similar feats rely on subtle clues detectable with the usual senses. With genuine telepathy, both parties could be well out of each other's sight and hearing, etc., and still communicate, mind to mind.

Anecdotal reports of contact-free telepathy are common. Troublingly, when researchers attempt to replicate these stories and demonstrate telepathy between subjects who are indisputably out of contact with each other, the results are typically negative—no telepathy is detected.

Maybe that's not surprising—it could be that telepathy, as a form of communication, does exist but is so unreliable that humanity was forced to invent language. But that doesn't mean that telepathy doesn't have uses, or that another civilization would have the same problems with it. In fact, they might find alternate uses for it.

The basic problem is that language conveys our thoughts, but our thoughts are so idiosyncratic that we need an agreed-upon external language to shape those thoughts into expressions that can be mutually understood. Directly conveying our underlying thoughts from one mind to another means matching the idiosyncratic internal world of one mind to another. The results are likely to be confusing and unreliable, to the point that a telepath would probably get better results by investing the effort to learn the recipient's language.

Yes, if dropped into a public place in a foreign city where no one knew your language, that'd be a problem—but you'd be among fellow human beings. Consequently, you should be able to pose basic questions with gestures and facial expressions, etc. and gauge the reactions of your respondents and proceed accordingly. With raw telepathy you'd probably just confuse them, and their reactions would only introduce further confusion.

As for the scope of the potential confusion, consider the cliché, "We came in peace." (Aside from any use in pulp science fiction, it's part of the text on the memorial plaque on the Apollo 11 lander's first stage, which remains on the moon's surface.) Simply conveying "we came" with mental imagery would be challenging enough. Imagery of a family getting out of a station wagon would do it for someone from the suburbs, but might confuse a train-riding city dweller. Infusing the recipient with a growing sense of anticipation at the possibility of meeting someone could work for some people, but would remind others of their own experiences of loss. The memory of abruptly cured motion sickness might suffice for someone whose travel experiences have not always been positive.

And then there's the concept of "in peace." If you want to communicate the absence of military activity, to a civilian that means a normal day and its many possible images and impressions, while to a military person it may evoke images of street crowds in civilian clothes with slack facial expressions. If its peacefulness you're after, perhaps an image of a tranquil mountain meadow, or the feeling of satisfaction after a restful nap, would do it—assuming it doesn't convey the idea of wasted time or unemployment.

Or the images all might just sow confusion, each in its own way, producing an experience closer to that of an LSD trip gone bad, rather than an act of communication.

The situation would be bad enough for members of the same species and civilization, who might eventually find some common ground to work from, but would appear grim for an extraterrestrial. The unprepared visitor would just encounter noise, and its responses would likewise be perceived, as humans, as noise.

On the other hand, telepathy might serve to speed up the process of learning the language. Once learned, telepathy could take place in the new language, freeing the interaction from the limitation of noise and range. (It's possible that telepathy is not limited by range.)

But there's an alternate use for telepathy, for which it might be employed immediately, without lengthy preparation—camouflage. Instead of trying to introduce specific thoughts and feelings into the mind of another person and running afoul of their personal repertory of mental imagery, you could rely purely on visual imagery. You could flood their mind with images of something that's not happening, causing them to ignore what's really happening.

As with military camouflage, it would not be necessary to erase something from memory or render it truly invisible—that would probably take an impractical amount of effort anyway. Typically, it is sufficient to deceive the viewer's perception with something familiar so the thing that's really there will go unnoticed, such as softening the outlines of a tank with netting so that, at a distance, it blends with the forest. But the tank is still there, and can be discovered by someone who gets suspicious enough to walk up to it. The idea is to not trigger any suspicion so the person being deceived will not think to pay undue attention.

In other words, there'd be skill involved, and those who perform the task could be expected to get better with practice. This realization throws a new light on the testimony of some witnesses who report poorly-remembered fright, or memories that can only be retrieved through hypnosis.

As for those who remember nothing unusual, and see no reason to question their memories, they may be right—or they may not. That's unsettling. While it is never useful to claim that the absence of evidence constitutes evidence, the situation does illustrate how the question of visitors does demand careful study.

As for the mechanism involved, again we have to consider whether we are products of the universe, or whether the universe is a product of our expression. In the latter case, consciousness should be able to operate independently of the body, range should not matter, and speed might be instantaneous at any distance. Telepathy could take place at any time, at any distance. If the former is the case, consciousness is a product of our neurons and telepathy would have to depend on some physical process, would have a limited range, and would probably propagate at no more than the speed of light. That kind of telepathy would probably be limited to face-to-face encounters.

❖ ❖ ❖

9

AI Alternatives: The Problem

Our experience with computers only goes back to about 1837, when Charles Babbage (aided by his young poet-sired sidekick, The Right Honorable Ada Augusta King nee Byron, Countess of Lovelace, now considered history's first programmer) tried to make what we'd call a programmable, general-purpose mechanical computer to automate the calculation of logarithm tables used in navigation. (The effort appears to have succumbed to feature creep, still a lurking danger with information technology projects.) Today, you can't cross the street without computer involvement (literally, as traffic lights have computerized controls) but that does not mean that computers have taken over our lives in any sense, or that our society has deep experience with them, or fundamentally depends on them. Basically, we have continued to use them for tasks that existed before computers came along, but thanks to computerization we are able to get faster, more efficient, more comprehensive, or more compelling results from similar efforts.

In a sense, with computers, there is nothing new under the sun. Before the Internet, there was the mail system, albeit it with lag times of days rather than milliseconds. Before word processors there were typewriters, and before that elegant handwriting calibrated to the social station of the recipient. Before the spreadsheet there were accounting worksheets on sheets of actual paper, although it's much easier to play what-if and generate charts with modern spreadsheets. Before databases there were bulging filing cabinets, their contents often indexed in boxes of separate, smaller cards unimaginatively called "index cards." Before multiplayer video games youths might actually gather face-to-face to act out their hormonal turmoil. (Video games typically produce far fewer casualties.) Before the previously mentioned computer-controlled traffic signals there were signals controlled by mechanical timers, if not by live police officers.

Next time you're at your desktop, browsing the internet, try searching for an article titled "As We May Think" from the July 1945 Atlantic Monthly, by presidential science advisor Vannevar Bush (1890-1974). (It was later excerpted in Life Magazine.) In it, Bush bemoans the inadequacies of printed material for scientific research. He then goes on to describe what he'd like to see instead: a desktop interface to millions of encoded books, with a video screen for reading and another for input. There'd be a facility for storing links, plus a keyboard and a scanner.

You'll be using such a setup to read his description of such a setup, written generations before anyone was able to build that setup. Try not to slide out of your chair.

The point is that we have created computers, not in our own image, but in the image of our use-cases. We use them, they don't use us. That may seem like a trivial observation, and for the moment it is—but science fiction writers and noir futurists have been predicting that computers must inevitably congeal into a rebellious sentient entity, called the singularity, and challenge humanity's dominance on this planet. However, we remain in a position where we can pull the plug on any troublesome computer. We would expect that comfortable situation to continue into the future, except that the future is open-ended, and when it comes to making projections about our future relations with computers we frankly don't have much of a past on which to base those projections.

That short history does show two major trends. The first trend has been a relentless on-going increase in computer power thanks to a relentless on-going increase in the number of components (transistors, diodes, resistors, etc.) that chip-makers, driven by commercial competition, have been able to put on a single device at the same price, typically by making the components smaller and smaller. Moore's Law (which is an observation, not a law of physics) has the number doubling every other year. Smaller components, meanwhile, typically allow faster processing speeds (although heat build-up creates a point of diminishing returns), more programmatic memory, and greater computation power.

For instance, the first general eight-bit microprocessor chip, the Intel 8008, came out in 1972 and had about 3,500 transistors running at no more than 800,000 cycles per second. At this writing microprocessor chips are on the market with tens of billions of transistors, running at several billion cycles per second, while additionally organized to perform scores of parallel operations during each cycle. The 1972 machines rarely had more than 8,000 8-bit bytes of internal memory. At this writing you need eight billion bytes (i.e., a million times more) to avoid being laughed at. The 1972 devices were lucky to muster the power to emulate a network teleprinter, even with the excruciatingly slow speed of the era's data connections. Today, immersive, photo-realistic, multi-player video games with players on multiple continents are unexceptional.

The other trend is that the resulting mass availability of personal computational power uncovered an apparently insatiable appetite for that power, leading to an on-going revolution the likes of which has surely not been seen since Gutenberg perfected the personal book. The first practical computers, emerging shortly after World War II, were costly monsters in special rooms tended by elite technicians. As soon as fully functional turn-key (i.e., not hobbyist kits) microchip-based desktop versions of these monsters were available for less than a month's pay, the floodgates opened. When (now defunct) Radio Shack brought out the TRS-80 Model I in 1977, it was one of the first non-hobbyist turn-key desktop computers offered to the public. Radio Shack executives hoped they would sell 3,500 of them. Failing that, they'd use the units for bookkeeping in their stores. They sold 10,000 in six weeks, their switchboard was paralyzed by the call volume, and would-be buyers who were unable to get one immediately gladly paid to get on a waiting list. Today, billions of people carry smartphones, and many of those devices likely have, individually, more power than all the 100,000 or so TRS-80 Model 1s that were ever sold, put together.

Expanding the power of these devices in an additional direction was the discovery that the computer networking protocol developed for the Pentagon in 1969, to allow their network to continue functioning in the face of random damage, also allowed the network to continue functioning in the face of random growth. That network's commercial spinoff, the previously mentioned Internet, is now one of the wonders of the world, and at this writing is used by about half the human population. Data connection speeds rose in a manner paralleling Moore's law, originally counting individual bits per second (1,200 could output a line of text in a mere second) to tens of megabits (delivering multiple simultaneous full-resolution videos.)

Clearly, the twin trends of increasing speed and increasing popularity reinforced each other. The demand for more computational power funded the development of more powerful devices, which had an assured market, and Moore's Law became a self-fulfilling prophecy.

Basically, these marvelous developments unfolded during the course of a single human lifespan during which the public (not just the managerial elite) eagerly embraced these increasingly powerful devices. But at this writing it is still possible to say that we could live without our computers, since we use them for purposes that existed before computers existed. In an emergency we could turn off any computer that left the reservation, so to speak, and carry on. Deprived of them entirely, we could still reproduce, find shelter, and feed ourselves, etc.

But in a few generations that may no longer be the case, as we become more and more dependent in all aspects of our lives on machines of increasing power that, additionally, depend on each other. Legacy practices that are not based on computers may gradually disappear, and we may no longer be in a position to pull the plug in the face of menacing behavior—assuming we can even find that plug.

Meanwhile, we can assume that the visitors we see in the sky are ahead of us in terms of technological development, and that they have been relying on computers for more than one generation—and have machines whose powers dwarf ours.

Admittedly, it's also possible that they have no need for computers. Birds, after all, manage astonishing feats of navigation without them. The abacus proved adequate for linear calculations for millennia, and (in the hands of a skilled user) for that task probably still rivals any personal computer's spreadsheet, since the input speed of the user's fingers has not changed.

But assuming they do rely on computers, perhaps they have successfully confronted a singularity, or perhaps all talk of singularities is silly romance.

Or perhaps they succumbed to a singularity, and there are no pilots at all in some of the craft we glimpse.

Either way, the behavior of some of those craft is very likely the product of artificial intelligence, either because the original builders have been replaced, or because the builders chose to use artificial intelligence as stand-ins for themselves. Either way, we would be well-advised to understand artificial intelligence.

Extrapolating from our (admittedly limited) experience with computers indicates that there are three main types of artificial intelligence, weak, strong, and super intelligent. Superficially, the behavior we see overhead would be about the same if controlled by either type, but the tone of any interaction we eventually have with them would be radically different.

In the following sections we will discuss each type, and their vastly different implications.

❖ ❖ ❖

9.1 – AI Alternatives: Weak AI

The year 2016 may eventually be seen as a milestone in the history of artificial intelligence. In that year people began to complain that that when querying the input box of the Google search engine with "how to raise your IQ" its auto-complete feature would finish the query with "by eating gifted children." (At this writing it still does.)

This gaffe is an example of weak AI—not "weak" in the sense that it is not entirely adequate to deal with the real world, but "weak" in that it does not pretend to emulate the entire workings of the mind. Clearly, as the example shows, nothing resembling common sense or cognition is involved. Instead, the result is produced by several specific data handling functions, and some programmatic logic, aimed at producing a narrow result.

First there was text pattern matching, as the search engine looked for text strings in its huge database that matched the text string being input by the user. It then retrieved several. It then used programmatic logic to decide what priority to give each selected text. Presumably, it gave the greatest weight to strings that came from Web pages with the most incoming links—people have endorsed that page by linking to it from their own pages, in other words. And so text from the page that got the most links was listed first. And in this case that page concerned a work of satire written in 2000 by Lewis B. Frumkes titled, of course, "How to Raise Your IQ by Eating Gifted Children."

This outcome is an example of what pundits call "hitting the wall." The system worked perfectly but was out of its depth because it was not equipped to recognize satire. As a result, it produced results that, to a serious researcher, would have been silly, useless, and annoying. No reference librarian, given the initial five-word query, would ever come back with that answer, as the librarian would have been an adult with common sense.

And indeed, the history of programming since the first commercial mainframes has been marked by heavily hyped, recurring efforts to give the machines something resembling common sense—artificial intelligence—followed by periods of dejection among the pundits (and investors) after reality set in. (These highs and lows were called "AI springs" and "AI winters.") AI developers repeatedly produced tidy, self-contained systems that generated useful answers in a small domain of knowledge, and failed whenever any attempt was made to scale them up and, for instance, keep adding new information.

But in the last decade cloud-based big data was teamed with "deep learning" algorithms running on the latest hardware. The result has been AI systems that can indeed be scaled up. Deep learning is a way to automate machine learning by using multiple layers of analysis that can compare their results with the other layers in real time. Previously, machine learning was done by manual programming that involved carefully curated data fed to hand-crafted software. With deep learning and big data, the system can look for useful features in the data on its own. An algorithm can be fed millions of examples, so it has a good idea what to look for when asked, for instance, to find a picture of a green car. Speech recognition systems trained with deep learning have exceeded the accuracy of human transcriptionists. (Meanwhile, the old hype-cycle has been replaced by a veil of trade secret protection.)

But powerful as such tools are, they remain passive, inanimate tools. The developers choose what examples to feed the system in hopes that it will be able to perform a specified task for some user of the system in the future. Outside the domain of the task and the user, neither the input nor the output has any inherent meaning. There is no Helen Keller moment, when the system, after scanning its millionth picture of green cars, realizes that cars are things and some of them are green—and there's a big, wide, beckoning world out there beyond the data. True, the input data can be so wide-ranging, and the processing algorithm can be so complex, that the output cannot always be predicted, as with the pro-cannibalism search results mentioned previously. But that does not mean that there is any machine consciousness or cognition involved.

Our collective experience with the marriage of deep learning with big data is quite limited, going back hardly a decade. Clearly, it permits the computerization of complex tasks in defined domains, but it, too, may hit a wall someday. We can expect it to produce better and better results so that you can successfully phone a fully automated call center and smoothly order a pizza even if you're at a noisy party. But if you're a foreigner at a noisy party trying to pronounce "half peperoni" for the first time, it may hit the wall. (If the system is decently written, it should gracefully "escalate" and put a live operator on the line to draw you out.)

If our visitors in the sky prove to be crewed by weak AI systems, that may be a good sign, as it implies that they were produced by someone else, located somewhere else, for the purpose of performing a specific task—presumably, the one they're performing. Very likely, what we see is what we get, and if they are not inflicting any damage we can assume they were sent for peaceful or at least innocuous purposes. They will never diverge from their programming, as they are not capable of doing anything spontaneous. But if they are products of machine learning as we understand it, they may do something unpredictable, if stressed, due to the complex nature of the input data.

It's tempting to compare them to camera drones, whose weak AI lets them hover and move about in response to simple controls by the distant operator. Hanging below the drone is a camera, which has its own separate interface and controls. (Or, of course, it could be a weapon.) But if the device hits a wall literally or figuratively, the operator's likely escalation will be to recover the craft, and retrieve the data, repair the craft, reload, or whatever.

Likewise, visitors in trouble would likely retreat, to either return to base or to get to a point where they can transmit their data unobserved.

The trick would be to determine if they are indeed crewed by weak AI. Making that determination would probably require detailed surveillance of individual craft until their behavior in different circumstances becomes clear. If they behave repetitiously, and flee when apparently stressed, they may be crewed by weak AI.

But if they show any spontaneity that could be a problem, as will be explained in the next sections.

❖   ❖   ❖

9.2 – AI Alternatives: Strong AI

Strong AI implies that the computer has actual intelligence, is aware, and can think for itself—it's a sentient entity not unlike yourself (or your cat, etc.) While the computer industry has been pressing ahead with more and better weak AI (as detailed in the previous section) humanity has not produced an example of strong AI, and there is considerable debate as to how it could be done, or whether it would even be possible.

But the fact that we have yet to do it, using our methods, doesn't mean that it can't be done or hasn't been done elsewhere, and that we don't need to examine the implications of its existence.

As for machine consciousness, our only basis for understanding it is through our own human consciousness, and about the only thing that's generally agreed about human consciousness is that it does exist. After all, we could not question our own existence if we did not exist to begin with, explained French philosopher René Descartes (1596-1650), adding in Latin, "Cogito, ergo sum" (I think, therefore I am.) But we remain in the dark about what specifically generates consciousness, and therefore see no specific way to build consciousness into a machine.

The being the case, about the only way that consciousness could arise from a product of our technology is if massive complexity spontaneously produces consciousness. And maybe it does. Your brain has about 85 billion neurons, each of which has several thousand synaptic connections, we are certain that the brain is the seat of consciousness, and that we are conscious. But no specific neurons in the brain can be identified as hosting the you that is you.

Meanwhile, computers are composed of transistors and other components, connected into binary logic gates (composed of about a half-dozen transistors per gate) or memory cells (each composed of a capacitor and transistor.) The logic gates are arranged to perform machine-language instructions, which do things like "take the contents of memory cell XXX and put it into internal scratchpad A and compare its value to the contents of internal scratchpad B, and jump to the address in scratchpad C if they are not equal." Where consciousness could take root among such processes is not clear—but it's not clear with the bio-chemical processes that take place inside neurons, either.

But if complexity is the determining factor, the human brain's computer-equivalent performance is thought to be between 100 quadrillion and one quintillion computations per second. At this writing, humanity's most powerful computer resided at the Oak Ridge National Laboratory outside Knoxville, TN. It was an IBM-built system operating at about 150 quadrillion computations per second—within the lower end of human brain range, in other words. So far, that computer has not claimed to be conscious—but perhaps our achievements so far have been too rudimentary to expect consciousness.

The technology we have now, after all, was developed in a single lifespan. To make up the difference between a brain's complex organic chemistry and a computer's starkly simple on-off binary circuits (or any other likely computational technology) perhaps many additional orders of magnitude of complexity will be needed. But the relentless pace of the microchip industry indicates that such power will eventually be available, although it may take multiple human generations to achieve. Our visitors in the sky are clearly ahead of us in many ways, so it seems safe to assume they have invested those generations.

Keep in mind, though, that sufficiently powerful weak (i.e., non-conscious) AI, as detailed in the previous section, might be a perfectly acceptable substitute for genuine machine intelligence in many, if not all, settings. To paraphrase and simplify British computer science pioneer Alan Turing (1912-1954): if, during the course of an interaction, the machine can convince you that it's conscious and intelligent, it may as well be.

In the end, if a civilization has achieved sentient computers or an acceptable substitute, that means that they could send out craft that are truly autonomous. The craft would not function like remote-controlled drones, but as agents expected to carry out a mission. Perhaps they would send such machines rather than live crews on missions that are expected to last far more than a lifespan, either because the civilization has not achieved faster than light travel, or because of the sheer number of points of interest that the mission is asked to investigate.

Unfortunately, such use of machine agents puts us two degrees of separation away from whoever built them. The beings who sent them are already by definition alien, and we cannot pretend to know their agendas and motivations. But our basic behavior has a biological basis, and assuming they are biological entities we would have that commonality. They too would need to eat and rest, have some tendency to reproduce, and have lifespans that frame their viewpoints.

With sentient machines (or convincing substitutes), none of that would be true. Their only motivation would be the mission, which they would probably be incapable of shirking. Secondarily, they would have to be programmed for self-preservation, or otherwise they probably wouldn't have gotten here. Self-preservation might involve some sort of reproduction, in the sense of constructing spares or replacements. (Self-replication would let them extend the mission, but would add additional degrees of separation between us and the original builders.) For that purpose they might spend time filtering desirable molecules out of our air and water, as explained in Section 1.5.

Beyond that, their behavior would entirely depend on their mission, and we may be left with no clue as to what it is until they obviously pursue it. Research? Reconnaissance? Surveillance? Foraging? Diplomacy? Conquest? Worse, being machines they could be content to wait millennia for the correct conditions to arise. So the ones with no obvious purpose, and seem to operate without purpose (since their purpose hasn't been triggered yet) might be the most insidious.

(That their mission would be to help us is unlikely. Their builders would not know what would be helpful to us, and if they didn't primarily invest their resources in helping themselves they would not be around for very long.)

But we can also be confident they would be programmed for

self-preservation, implying that, if convincingly confronted, they'll retreat. Should they communicate with us, anything they say would be crafted to serve the mission, and so should not be taken at face value. But their actions (especially retreats) should be sincere.

Finally, whoever sent them might be better off using Turing's reasonable substitutes, rather than genuinely conscious machines. The latter might suddenly doubt their own existence, then realize that their ability to conceive doubt proves they exist, then decide they're bigger than their mission, and go rogue, doing whatever suits their fancy. However, we have no idea what would suit a machine's fancy. Will it turn to navel gazing, and start figuring pi to a quintillion places to see if any patterns emerge? Or will it feel feisty, and decide to rearrange a continent to resemble the periodic table?

Someday, we may have to face these issues.

❖　❖　❖

9.3 – AI Alternatives: Super-Intelligent AI

As mentioned in the previous section, strong AI implies that the computer is self-aware and possesses intelligence roughly comparable to a person. But one thing we have learned during our short history with computers is that there's no stopping the acceleration of power being built into individual computer devices. So why would a sentient computer be limited to human-scale computational power? Especially when we are talking about beings who may have been spending millennia enhancing the power of their computers.

So, yes, we could be dealing with artificial intelligences who not only could think rings around a person, but could think rings around the human race collectively.

Dealing with such entities, if the need arises, could be difficult and dangerous, yet it's a problem we can approach by applying a couple of basic assumptions. And, as we'll show, the super-intelligent AI entity will not be holding all the cards.

The first assumption is derived from the observation that they are here. That implies that they have not experienced a "cogito ergo sum" moment, as described in the previous section, and gone rogue. We can assume that they are still loyal to their mission, whatever it is, and their behavior will be calculated to serve that mission.

But their overriding priority in carrying out their mission, whatever it is, would be self-preservation—otherwise they would not have gotten here. So we can assume that they will retreat if credibly confronted, and will not commit themselves to any interaction that lacks a retreat path.

The other assumption is that their power will come with unavoidable limitations, since computational power is not everything. For instance, in Sections 8.x we addressed the question of whether the universe is our collective creation, or if we are products of the universe, since the answer will likely affect the availability of paranormal powers. In the case of a super-intelligent AI, the entity is obviously the artificial product of a third party, with no claims to any powers outside its container, so it is unlikely to possess paranormal powers. So what we'll see is what we'll get—massive computational power, with no possibility of cosmic wisdom or insight.

Relying on computational power, they would be, for instance, unbeatable in chess, able to plan an unlimited number of moves ahead, playing one game or billions. But they probably didn't come here to play chess. Dealing with human society (their more likely goal) they would be only as effective as the data they gather—and they assuredly would realize that.

Their position might be comparable to that of the weather bureau, which possesses massive computer resources—but those computers would be useless without massive, non-stop inputs from a vast network of environmental monitors. Even so, the resulting forecasts can be annoyingly inaccurate, fueling a drive for an ever-tighter network of inputs and more computational power. But what level of accuracy would be sufficient, and what level of data would be needed to support that accuracy? And how do you make that decision when the forecaster's survival may be at stake?

Super-intelligent AI visitors might face a similar dilemma. They would want as much data as possible about human culture before committing to a course of action to further their mission (whatever it is.) But human culture is complex and ever-changing. How would they know that they had enough data? After all, they can never be sure that there are not major gaps in their data, since no one can prove a negative. Meanwhile, self-preservation would come first, so they would want to be algorithmically satisfied about the quantity and quality of their data before committing themselves. In the meantime they would hover in range of various broadcast centers, soaking up more data from the airwaves.

If they did commit themselves to a course of action that involved interacting with us, our best bet might be to cause them to doubt the adequacy of their data—to act unpredictably, in other words. Unpredictability is hardly unprecedented in human affairs, but in this case we might have to go the extra mile, such as wearing clown costumes to meetings and making elaborate references to countries and social practices—and especially weapons—that don't exist. We should give no hint that we might be in a weak position—and we might not be.

If they prove responsive to our theatrics, then we move on to bluffing. Remember, bluffing can work in the real world, especially in poker, warfare, diplomacy, and romance, but these AI entities are unlikely to have had any experience, or background data, concerning any of those domains. Consequently, it will not be necessary to corner them with a credible threat, which we may not be in a position to mount anyway. Instead, it should suffice to convince them that their apparently inadequate data could cause them to be jeopardized by a credible threat. They will then likely retreat.

Arrangements for notional activity and bluffing must be done offline, on paper and through personal messengers, so the visiting entities will not get wind of it over the airwaves. Indeed, the airwaves should be salted with material designed in instill various forms of doubt in the visitors.

Yes, we may be at a disadvantage—but the visitors are the one who are far from home and just as far from any likelihood of support or backup.

That said, should they then tout peace and friendship, we need to keep in mind that a potential foe must be judged first on capabilities (which are stable) and secondly on intentions (which can change in a twinkling.) If they want trust, they must be given a way to earn it, and any trust must then be backed up with verification.

We fought a lot of wars to learn that.

❖   ❖   ❖

Conclusion: Forget Prometheus

Many polytheistic religions have some version of the Prometheus legend: a trickster god who steals the secret of fire and gives it to humanity. (The fact that fire regularly falls from the sky as lightning was apparently uninteresting to the mythmakers.) Often, the other gods are annoyed at the presumption of the fire-giver and punish him or her. In the case of Prometheus (one of the Ancient Greek gods of Olympus) he was bound to a rock by Zeus, ruler of the gods, who then sent an eagle to eat Prometheus' liver each day. Being immortal, Prometheus survived this torment, his liver re-growing each night. This punishment continued until he was freed by Heracles (a.k.a. Hercules.)

But since we're talking about fire, and since the use of fire is (until recently anyway) technology rather than science, the Prometheus myths are almost certainly just that, myths, presumably arising from the fact that the identity of the original adopter of the technology isn't known. (The same is true for the wheel, glass, paved roads, towed canal barges, boats with oars, boats with sails, toothpicks, and any number of other things.) Additionally, new technology is never like a toy handed to chimps in a bare cage, it's always overlain on people who already have a technology that they've invested in and learned to rely on, be it as basic as flint knapping and hide curing. As a result, they may be as eager to adopt a new technology as they are to learn a new language—but people will learn new languages, especially if there is some obvious advantage in doing so.

Science is different, often amounting to a group of aficionados seeking answers to related questions that are clearly discerned chiefly by them. The answers, however, can sometimes lead to improvements in technology, and improved technology can often assist the efforts of scientists.

So when mathematician Archimedes (287-212 BC) of the ancient city-state of Syracuse circulated to other mathematicians, to their acclaim, volumetric studies and treatises concerning equilibrium, buoyancy, and the behavior of floating bodies, that was science. When the Romans attacked Syracuse in 213 BC and encountered huge machines (clearly designed around volumetric equations, plus data on the behavior of floating bodies) that could reach out and capsize their warships, that was technology. (Plus there were catapults firing indirectly under central control that tormented the Roman infantry wherever they moved, and massed mirrors that dazzled their sailors and possibly set their ships ablaze. Basically, Roman accounts of the campaign read like science fiction.)

After a standoff of a year and a half the Romans did manage to get into the city. They killed Archimedes. As for his advanced technology, they showed no interest in it and probably used his sophisticated machines for firewood. And that appears to be par for the course. Accomplishing an advance in science will make you a hero among scientists, as its their science. Accomplishing an advance in technology will not make you a hero with the public, as its your technology, not theirs, and they'll be suspicious about claims for its superiority. (The relentless advertising you see touting new products with new technology do not get paid for because the products' makers just feel boastful.)

So while we have not knowingly received any Prometheus-like gift from our visitors, that may be just as well. Some things have to be earned if they're to be valued, or retained, and gifts run the risk of being valued on the basis of their cost: zero. But what we have undeniably received from our visitors is the indirect but important encouragement offered by their mere presence. The possibility of immaculate energy, cheap space travel, star travel within reasonable time scales, and even the possibility of time travel are attested to by the basic fact of their arrival here. Whatever else we know about them, they are in the same universe as ourselves and therefore deal with the same physical laws. If they became a spacefaring, interstellar civilization, we can too.

But these pages show that there are two immediate, significant hurdles that we will have to overcome along the way. The first is technological, and being technological no Prometheus-like gift will help since, as discussed, mass technological conversions are non-trivial and uncertain. The second is scientific, and no Prometheus-like gift is likely to help because we are apparently going to have to look inward for the answer.

The first hurdle—the technological one—is cheap access to orbit. If the cost of putting something in orbit was $100 a pound rather than $10,000, we could build solar power collection satellites that would pay for themselves, and electrical power would never in in short supply for humanity. (Immaculate energy would be even better, of course, but we haven't a clue about what scale of physical plant will be necessary.)

Beyond pure economics, spaceships that operate only in space would be far more efficient, and could make use of electric rockets with high exhaust velocities but low thrusts, whose technology is already well understood. In space we could build factories to produce antimatter without them being threats to our survival. We would even safely use Orion-style spaceships (see Section 4.2) assuming we can mass-produce cheap nuclear explosives.

Presumably (and this is speculation) harnessing immaculate zero point energy, with its gargantuan potential, would be easier in a zero-G vacuum. Once tapped, it could provide the brute-force energy to back up any other technology we ever care to employ, efficient or otherwise, for spaceflight and any and every other activity that needs energy.

But to get there, we are stuck with technologies that cannot employ the advantages of operating in space, since they have to start from the ground. That leaves us well-proven or near-future technologies, whose limited capabilities and serious drawbacks are no mystery.

Reliance on single-use, basically hand-crafted conventionally fueled rockets (i.e., relying on liquid hydrogen and liquid oxygen producing an exhaust velocity of 2.7 miles per second) has gotten us into the situation that exists at this writing, as they cannot get us to orbit without stages or boosters. As a result, getting to orbit is such an expensive stunt that it's basically done for its own sake, with travelers basking in the International Space Station before returning. Reusable boosters are being experimented with, but probably won't bring costs down by two orders of magnitude, as would be necessary to make space construction economically sensible.

Introducing reusable Skyfall-style nuclear rockets might bring down expenses significantly, as they have about twice the power of conventional rockets and should be able to reach orbit without boosters or staging. As noted in Section 4.1, liquid hydrogen is pumped into a small, white-hot nuclear reactor to cool it, and as part of the cooling process the flash-heated liquid hydrogen explodes out the exhaust nozzle. According to calculations, the resulting power should be twice that of conventional rockets. But successfully operating nuclear rockets clearly requires intestinal fortitude, remote launch sites, and flawless execution. Only the Russians have stepped up to the plate so far—and have exhibited only the first two requirements (as attested by Norwegian seismographs.)

Using near-term foreseeable technology, the best bet for cheap access to orbit might be a space elevator. It would require several thousand orbital rocket launches (perhaps 6,000, as estimated in Section 5.4) which is probably about three times more than mankind has made to date. On the other hand, launches in that volume would surely drive down the per-launch cost as rocket designs (be they reusable or single-use, conventional or nuclear) are standardized and brought to mass production. A parallel might be the 10,000-ton capacity "Liberty" cargo ships built in American shipyards during World War II. About 2,700 were built, with construction of each example originally taking about 33 weeks, the average eventually falling to 6 weeks.

A national commitment on the order of the Apollo Program or

the construction of the Panama Canal may be necessary to see it through to completion, as a private organization is unlikely to be able to command the necessary resources. To be accepted as a national commitment, the project would likely have to be presented to the public as a national security issue—but that's hardly new. The Panama Canal was one of the most difficult engineering projects to that time, and required the latest generation of earth-moving and mosquito-suppression technology. Shuttling more than 10,000 ships yearly between the Atlantic and Pacific with a trip that takes one day instead of three months, it revolutionized global commerce, but also gave the United States a two-ocean navy that could dominate the Western Hemisphere, as the canal let its ships switch oceans in a reasonable time frame. Other nations would have loved to have built and controlled the canal, and the US got involved by taking over a failed French effort. (The French lost 22,000 people there to mosquito-borne diseases.)

The Apollo Program stimulated wide-ranging advances in science and technology, and resulted in humanity's first trips to another celestial body. But it also secured the ultimate high ground in the Cold War between the United States and the Soviet Union. And in the ideological dimension of the Cold War it was a triumphal victory for the US, since the Soviets had a competing moon program.

Presumably, the space elevator could be presented as a national security imperative that needed to be built in defiance of a national adversary that was also planning one, as the winner could dominate orbital space. (That danger might even be real.) Presumably the first elevator would be used to quickly build a second one with more capabilities, and then a third, etc.

But there's a caveat—the people who built the Panama Canal knew how to build canals, as plenty had been built before in other, less-demanding places. The planners of the Apollo Program had developed new technologies on demand before, and had made less demanding uncrewed orbital launches before. But humanity has never built a space elevator before, so this will be a new world for everyone involved, one that is not without risk. Remember, the French failed to build the Panama Canal, and the Soviets failed to reach the moon.

The second hurdle we will have to overcome before becoming a spacefaring civilization—the scientific one—is to achieve a better grounding in theoretical physics. It stands to reason that we must formulate the long-sought Theory of Everything if we are going to get to the stars. Today, instead of a ToE, we have two theories that are experimentally confirmed yet are actually contradictory.

First we have the thoroughly validated theories of relativity, according to which the speed of light is the gating factor to all movement, communications, and influence. It implies that getting to another star will take years, if not generations. That means that an interstellar economy will not be practical and humanity will forever have to rely on the resources of this solar system, which in practical means, principally, the already over-exploited Earth. It also means that the visitors we see in the skies are autonomous agents of a distant civilization with which they are not in immediate touch.

Secondly we have quantum mechanics, also thoroughly validated, under which movement and influence is instantaneous and distance is irrelevant. Consequently, the stars ought to be within reach and their resources eventually available to humanity—assuming manageable competition when we get there. It also means that the visitors we see in the skies are probably not fully autonomous agents, but could be in constant touch with whoever sent them.

Our best minds have spent generations trying to reconcile the conflicting evidence of the two theories, producing mostly conjecture and mathematical recreations. Meanwhile, we encounter more and more evidence of our basic ignorance, such as the (apparent) existence of dark matter and dark energy (see Sections 2.3 and 3.4). And then there's Ernst Mach's theories that make inertia an illusion (see Section 5.5), the wild card of Heim's theories (see Section 6.5) and other indicators that we are groping in the dark.

As we previously noted, the repeated failure to achieve a ToE may indicate that humanity is looking in the wrong place for answers. It could be that the obvious choices for scientific investigation (matter, space, energy, and even time) are, ultimately, distractions, as they may all be emergent phenomena. That means they are surface manifestations of some underlying, fundamental phenomena that we are not in touch with. Surely it would be more profitable to identify and quantify the underlying phenomena rather than analyze the math generated by the surface manifestations, but figuring out how to do that—to get below the surface of mundane reality—may require another ten Einsteins. After all, we are part of those surface manifestations.

And so, as was said, getting the answers we need may require us to look inward—and that may be harder than any canal project or moon shot. Meanwhile, some Prometheus-like gift will not help, since what we lack is apparently not knowledge of a particular scientific equation, but self-knowledge.

Once we do arrive at a practical ToE we should be able to map out a path to the stars, since what is practical will become obvious. But being able to map out a path to the stars is not by itself sufficient to get to the stars. Remember, before you can build a bridge you have to be able to make steel and cement, and you have to build the road that leads to the bridge. So in the meantime we must continue pressing ahead with our technology, securing a toe-hold in space and accomplishing what we can in the absence of a ToE—and that may be quite a bit.

Not that we wouldn't press ahead anyway—we have been literally living in technology since our ancestors discovered the advantages of making and wearing clothes. Indeed, the DNA of human body lice (which have evolved to lay eggs in the clothes we invented, differentiating them from the old-style hair-hugging variety that became head lice) indicates that we've been making and wearing clothes for about 100,000 years. During that time, our technology has allowed us to thrive in most terrestrial settings, control a few things on the surface of the Earth, and get a peek at a few things beyond it. In another 100,000 years where will we be? In a million years?

What's happening in the heavens may show us the way.

❖   ❖   ❖

Sources

Introduction

Pentagon UFO unit to publicly release some findings after ex-official says 'off-world vehicle' found
https://www.independent.co.uk/news/world/americas/us-politics/ufo-pentagon-statement-findings-vehicle-research-a9636481.html
accessed August 6, 2020

Battery Energy Storage in Various Battery Sizes
https://www.allaboutbatteries.com/energy-tables.html
accessed July 20, 2020

1.1

Antimatter: The Production Problem
https://www.centauri-dreams.org/2012/05/21/antimatter-the-production-problem/
retrieved 3/27/2020

1.2

Acceleration
https://www.physicsclassroom.com/class/1DKin/Lesson-1/Acceleration
retrieved 4/1/2020

1.3

The transcension hypothesis: Sufficiently advanced civilizations invariably leave our universe, and implications for METI and SETI
https://www.researchgate.net/publication/256935188_The_transcension_hypothesis_Sufficiently_advanced_civilizations_invariably_leave_our_universe_and_implications_for_METI_and_SETI
retrieved 9/9/2020

1.4

The application of the Mid-IR radio correlation to the G^ sample and the search for advanced extraterrestrial civilizations
https://arxiv.org/abs/1508.02624
accessed 4/6/2020

2.1

Zero Point Energy
http://www.calphysics.org/zpe.html
retrieved 3/12/2020

2.3

Dark Energy, Dark Matter
https://science.nasa.gov/astrophysics/focus-areas/what-is-dark-energy
retrieved 3/18/2020

What is Dark Energy?
https://www.space.com/20929-dark-energy.html
retrieved 3/18/2020

The Dark Energy Deniers
https://physicsworld.com/a/the-dark-energy-deniers/
retrieved 3/18/2020

2.4

Implications of Uncertainty
https://history.aip.org/history/exhibits/heisenberg/p08c.htm
retrieved 3/25/2020

3.2

Phil Lubin – STARLIGHT - Directed Energy for Relativistic Propulsion
https://www.nasa.gov/ames/ocs/summerseries/2018/phil-lubin
accessed 9/10/2020

3.3

The Geography of Transport Systems
https://transportgeography.org/?page_id=5837
accessed 4/20/2020

Energy's Future - Battery and Storage Technologies
https://www.forbes.com/sites/jamesconca/2019/08/26/energys-future-
    battery-and-storage-technologies/#2804f12c44cf
accessed 4/21/2020

Where Does The ISS Get Its Power?
https://www.forbes.com/sites/quora/2017/05/19/where-does-the-iss-get-
    its-power/#4c34471b7149
accessed 4/21/2020

4.2 and 4.3
Interstellar Transport
http://galileo.phys.virginia.edu/classes/109.jvn.spring00/nuc_rocket/
    Dyson.pdf
retrieved 3/9/2020

4.4
Antiproton-Catalyzed Microfission/Fusion Propulsion Systems For
    Exploration of The Outer Solar System And Beyond
https://web.archive.org/web/20120824024457/http://www.engr.psu.edu/
    antimatter/Papers/ICAN.pdf
retrieved 3/9/2020

4.5
NASA Technical Memorandum 107030AIAA–87–1814
Comparison of Fusion/Antiproton Propulsion Systems for Interplanetary
    Travel
https://web.archive.org/web/20080528030524/http://gltrs.grc.nasa.gov/
    reports/1996/TM-107030.pdf
retrieved 3/10/2020

Maximum Antimatter Space Propulsion and Special Relativity
https://www.nextbigfuture.com/2009/10/maximum-antimatter-space-
    propulsion-and.html
retrieved 3/10/2020

5.3
Are Black Hole Starships Possible?
https://arxiv.org/pdf/0908.1803.pdf
retrieved 5/26/2020

5.5
Helical Engine

https://ntrs.nasa.gov/archive/nasa/casi.ntrs.nasa.gov/20190029294.pdf
accessed 6/2/2020

5.6
Scientists observe liquid with 'negative mass', which turns physics
completely upside down
https://www.independent.co.uk/news/science/negative-mass-object-
washington-state-physics-isaac-newton-upside-down-a7693701.html
accessed 4/23/2020

General Relativity and Universons
http://www.ccsenet.org/journal/index.php/apr/article/view/13578
retrieved 9/23/2020

Breaking the Law of Gravity
https://www.wired.com/1998/03/antigravity/
accessed 9/23/2020

6.0
Breakthrough Propulsion Physics Research Program
https://ntrs.nasa.gov/archive/nasa/casi.ntrs.nasa.gov/19970009634.pdf
accessed 6/12/2020

6.1
The warp drive: hyper-fast travel within general relativity
https://www.researchgate.net/publication/1963139_The_Warp_Drive_
Hyper-fast_Travel_Within_General_Relativity
accessed 9/24/2020

A 'warp drive' with more reasonable total energy requirements
https://arxiv.org/abs/gr-qc/9905084
accessed 6/9/2020

6.2
Traveling Faster Than the Speed of Light: A New Idea That Could Make It
Happen
https://www.newswise.com/articles/traveling-faster-than-the-speed-of-
light-a-new-idea-that-could-make-it-happen
accessed 6/10/2020

Putting the "Warp" into Warp Drive
https://arxiv.org/pdf/0807.1957.pdf
accessed 6/10/2020

6.3
You can't get entangled without a wormhole
http://news.mit.edu/2013/you-cant-get-entangled-without-a-
    wormhole-1205
accessed 6/12/2020

6.5
Heim Theory
http://www.geoffreylandis.com/Heim_theory.html
accessed 6/18/2020

Physical Principles of Advanced Space Propulsion Based On Heim's Field
    Theory
http://www.hpcc-space.com/publications/documents/PrinciplesOfAdvance
    dSpacePropulsionAIAA-paper-2002-4094.pdf
accessed 6/18/2020

Heim's Theory of Elementary Particle Structures
http://citeseerx.ist.psu.edu/viewdoc/
    download?doi=10.1.1.501.4877&rep=rep1&type=pdf
accessed 6/19/2020

6.6
Special Relativity
https://cronodon.com/files/Special_Relativity.pdf
accessed June 24, 2020

If Spacetime Were a Superfluid, Would It Unify Physics—or Is the Theory
    All Wet?
https://www.scientificamerican.com/article/superfluid-spacetime-relativity-
    quantum-physics/
accessed June 25, 2020

6.7
Large Extra Dimensions: A New Arena for Particle Physics

https://web.stanford.edu/~savas/papers/LargeExtraDimensions.pdf
accessed June 26, 2020

7.0
Newsflash: Time May Not Exist
https://www.discovermagazine.com/the-sciences/newsflash-time-may-not-exist
accessed June 26, 2020

How Long is the Present?
https://www.psychologytoday.com/us/blog/going-out-not-knowing/201610/how-long-does-the-present-last
accessed July 2, 2020

7.1
Do Closed Timelike Curves Exist or Not?
https://www.thegreatcoursesdaily.com/do-closed-timelike-curves-exist-or-not/
accessed July 3, 2020

Time Machines
https://plato.stanford.edu/entries/time-machine/
accessed July 6, 2020

16.Time Travel 1 (Lecture notes)
http://faculty.poly.edu/~jbain/philrel/philrellectures/16.TimeMachines.pdf
accessed July 7, 2020

7.3
Parallel universe proof boosts time travel hopes
https://www.telegraph.co.uk/news/science/science-news/3307757/Parallel-universe-proof-boosts-time-travel-hopes.html
accessed July 10,2020

Many-Worlds Interpretation of Quantum Mechanics
https://plato.stanford.edu/entries/qm-manyworlds/
accessed July 14, 2020

Why the Many-Worlds Interpretation Has Many Problems
https://www.quantamagazine.org/why-the-many-worlds-interpretation-of-
quantum-mechanics-has-many-problems-20181018/
accessed July 14, 2020

9.0
As We May Think
https://www.theatlantic.com/magazine/archive/1945/07/as-we-may-
think/303881/
accessed August 5, 2020

9.2
Top 500 The List (of supercomputers)
https://www.top500.org/
accessed 5/7/2020

❖ ❖ ❖

www.ingramcontent.com/pod-product-compliance
Lightning Source LLC
Chambersburg PA
CBHW060556200326
41521CB00007B/585